UNIVERSITY OF BRADFORD
Library

BD 034064894 X

Springer Series in Applied Biology

Series Editor: Prof. Anthony W. Robards PhD, DSc, FIBiol

*Published titles:*

Foams: Physics, Chemistry and Structure

The 4-Quinolones: Antibacterial Agents in Vitro

Food Freezing: Today and Tomorrow

Biodegradation: Natural and Synthetic Materials

Immobilised Macromolecules: Application Potentials

# Human Health:
## The Contribution of Microorganisms

Edited by S. A. W. Gibson

Springer-Verlag
London Berlin Heidelberg New York
Paris Tokyo Hong Kong
Barcelona Budapest

Stewart A. W. Gibson
Reckitt and Colman Pharmaceuticals, Dansom Lane,
Kingston upon Hull HU8 7DS

*Series Editor*
Professor Anthony William Robards, BSc, PhD, DSc, DipRMS, FIBiol
Director, Institute for Applied Biology, Department of Biology,
University of York, York YO1 5DD, UK

304 200
UNIVERSITY OF BRADFORD
LIBRARY
- 1 SEP 1994
ACCESSION No. 0340648914 CLASS No. S 616·34·008·87 GIB
LOCATION

*Cover illustration:* Pathogenic *E. coli* adhered to gastric epithelium – courtesy of Dr Ashley Wilson, CCTR, Institute for Applied Biology, University of York

ISBN 3-540-19871-7 Springer-Verlag Berlin Heidelberg New York
ISBN 0-387-19871-7 Springer-Verlag New York Berlin Heidelberg

British Library Cataloguing in Publication Data
Human Health: Contribution of
Microorganisms. - (Springer Series in
Applied Biology)
I. Gibson, Stewart A. W. II. Series
576.162
ISBN 3-540-19871-7

Library of Congress Cataloging-in-Publication Data
A catalog record for this book is available from the Library of Congress

Apart from any fair dealing for the purposes of research or private study, or criticism or review, as permitted under the Copyright, Designs and Patents Act 1988, this publication may only be reproduced, stored or transmitted, in any form or by any means, with the prior permission in writing of the publishers, or in the case of reprographic reproduction in accordance with the terms of licences issued by the Copyright Licensing Agency. Enquiries concerning reproduction outside those terms should be sent to the publishers.

© Springer-Verlag London Limited 1994
Printed in Great Britain

The use of registered names, trademarks etc. in this publication does not imply, even in the absence of a specific statement, that such names are exempt from the relevant laws and regulations and therefore free for general use.

Product liability: The publisher can give no guarantee for information about drug dosage and application thereof contained in this book. In every individual case the respective user must check its accuracy by consulting other pharmaceutical literature.

Set by Institute for Applied Biology, Department of Biology, University of York
Printed and bound by the Athenæum Press Ltd, Newcastle upon Tyne
12/3830-543210 Printed on acid free paper

# Foreword from Series Editor

The Institute for Applied Biology was established by the Department of Biology at the University of York to consolidate and expand its existing activities in the field of applied biology. The Department of Biology at York contains a number of individual centres and groups specialising in particular areas of applied research which are associated with the Institute in providing a comprehensive facility for applied biology. Springer-Verlag has a long and successful history of publishing in the biosciences. The combination of these two forces leads to the "Springer Series in Applied Biology". The choice of subjects for seminars is made by our own editorial board and external sources who have identified the need for a particular topic to be addressed.

The first volume, *"Foams: Physics, Chemistry and Structure"*, has been followed by *"The 4-Quinolones: Antibacterial Agents in Vitro"*, *"Food Freezing: Today and Tomorrow"*, *"Biodegradation: Natural and Synthetic Materials"* and *"Immobilised Macromolecules: Application Potentials"*. The aim is to keep abreast of topics that have a special applied, and contemporary, interest. The current volume describes a group of non-pathogenic, host-derived microorganisms, probiotics, which may beneficially affect the host by improving the microbial balance of the target niche. The chapters describe the theories and evidence behind the concept of probiosis and examines the possible contribution of these microorganisms to human health.

The applications of Biology are fundamental to the continuing welfare of all people, whether by protecting their environment or by ensuring the health of their bodies. The objective of this series is to become an important means of disseminating the most up-to-date information in this field.

York, December 1993 A. W. Robards

# Editor's Preface

Probiotics could be defined as non-pathogenic, host derived microorganisms which beneficially affect the host by improving microbial balance of the target niche. This definition does not cover substances, such as lectins, which affect niche microbial balance chemically and may be considered narrow by those who wish to proffer non-viable microbes as probiotics. The use of these non-viable organisms and poorly contrived preparations in the early days of commercialisation in both the animal and human markets has bequeathed a bitter legacy which effective probiotics must now strive to overcome.

Modern scientific methodologies, properly formulated and tested products and rigorously controlled clinical trials have contributed to the partial re-establishment of probiotics. Many of the results are still inconclusive, however, and much work remains to be done.

This volume, containing contributions by internationally recognised experts from science and industry, examines the development of the human microflora from birth, its function in health and dysfunction in disease, the theories and evidence behind the concept of probiosis, the impact of probiotics on the microbial populations of various niches in the human body, clinical studies which are now being carried out to quantitate the probiotic effect, the steps required for the large scale production of commercially feasible preparations and the registration of these medicines with the MCA for use in humans.

With new European legislation (albeit slowly) set to target 'alternative' medicines, products will have to prove efficacy to the satisfaction of European authorities to survive. It is perhaps unfortunate that some potentially useful products, especially those from small or poorly financed organisations, will be lost during the ensuing purge. It is to the benefit of most companies and of the consumer that this process may help with the re-establishment of the health of probiotics in the minds (and bodies) of the scientific and medical communities and of end-users. Hopefully, this process will be completed before the now lengthy probation period for probiotics ends with the closing of the entry door to scientific and medical respectability.

S. A. W. Gibson

# Contents

# Contributors

Professor A. W. Bruce
Division of Urology, Department of Surgery, University of Toronto, Canada

Dr. P. L. Conway
University of Götebotg, Department of General and Marine Microbiology, Carl Skottsbergs Gata 22, S-413 19 Göteborg, Sweden

Dr. R. Fuller
59 Ryeish Green, Three Mile Cross, Reading RG7 1ES

Dr. G. R. Gibson
Medical Research Council, Dunn Clinical Nutrition Centre, 100 Tennis Court Road, Cambridge CB2 1QL

Dr S A W Gibson
Reckitt and Colman Pharmaceuticals, Dansom Lane, Kingston upon Hull HU8 7DS

Dr. M. A. Hall
The Princess Anne Hospital, Coxford Road, Southampton SO9 4HA

Dr. A. Henriksson
University of Götebotg, Department of General and Marine Microbiology, Carl Skottsbergs Gata 22, S-413 19 Göteborg, Sweden

Mr. S. Laulund
Chr. Hansen's Bio Systems A/S, Hørsholm, Denmark

Dr. A. Lidbeck
Department of Microbiology, Huddinge University Hospital, Karolinska Institute, Stockholm, Sweden

Dr. G. T. Macfarlane
Medical Research Council, Dunn Clinical Nutrition Centre, 100 Tennis Court Road, Cambridge CB2 1QL

Professor C. E. Nord
Department of Microbiology, Huddinge University Hospital, Karolinska Institute, Huddinge and National Bacteriological Laboratory, Stockholm, Sweden

Dr G. Reid
Department of Microbiology and Immunology, University of Western Ontario, SLB 328, London, Ontario, N6A 5B8 Canada and Division of Urology, Department of Surgery, University of Toronto, Canada

Ms. S. L. Smith
The Princess Anne Hospital, Coxford Road, Southampton SO9 4HA

Professor G. W. Tannock
Department of Microbiology, University of Otago, Dunedin, New Zealand

Ms. L. Tomeczek
Division of Urology, Department of Surgery, University of Toronto, Canada and Department of Microbiology, University of Toronto, Canada

Chapter 1

# The Acquisition of the Normal Microflora of the Gastrointestinal Tract

G. W. Tannock

## Introduction

The birth of a mammal must be a moment of celebration in the microbial world. A new, potential host emerges from the uterus where it has been protected during development from contact with bacteria, fungi and protozoa. During passage of the young animal through the birth canal, members of the vaginal microflora have the opportunity to contaminate the pristine surfaces of the infant. At the moment of birth, or shortly thereafter, microbes in faeces that have been involuntarily expelled by the mother during labour, and microbes that are present in the air or on inanimate materials enter the germfree ecosystems of the neonate and proliferate to a dramatic extent. Suckling, licking (kissing) and grooming (caressing) of the infant after birth enables transfer of skin and oral microbes from adult to neonate to occur. It might be thought, indeed, that the positioning of the orifice of the birth canal in close vicinity to the anus, and the nature of parental expressions of caring are of microbiological design, since they ensure the transmission of microbes comprising the normal microflora from one generation to the next (Carlsson and Gothefors 1975; Brunel and Gouet 1982; Brunel and Gouet 1989; Tannock *et al.* 1990b).

Colonisation of the neonatal surfaces and internal cavities that have openings to the body's exterior occurs within the 24 h following birth. Proliferation of microbial types in these sites appears to be initially unchecked, resulting in a heterogeneous collection of microbes. Soon, however, regulatory mechanisms generated within habitats (autogenic factors) and by external forces (allogenic factors) permit the continuing presence of some microbial types in the gastrointestinal ecosystem but the elimination of others. These qualitative and quantitative changes that occur in microbial populations inhabiting the young animal provide an example of a biological succession. Eventually, usually after weaning, the microbial composition of the microflora becomes more stable and the adult microflora (in ecological terms, the climax community) is attained. The microflora of adult animals is composed characteristically of numerous bacterial species, the numerically dominant of which are present at about $10^{10}$ bacteria per gram of intestinal sample (reviewed by Savage 1977 and Macfarlane and Gibson, chapter n this volume). Examples of all of these features can be observed during the acquisition of the gastrointestinal microflora by the laboratory mouse.

## Acquisition of the Gastrointestinal Microflora of Mice

The initial collection of microbes inhabiting the gastrointestinal tract of neonatal mice is composed of environmentally-derived bacteria (e.g. flavobacteria), facultatively anaerobic members of the normal microflora of the mother (staphylococci, enterococci, *Escherichia coli*), and aerotolerant bacteria with fermentative metabolism (lactobacilli) also of maternal source. Microbes originating in the general environment do not persist in the gastrointestinal tract for long, however, and by about ten days after birth, lactobacilli, enterococci and *E. coli* are especially numerous (about $10^8$ per gram of caecal contents). Lactobacilli predominate in the proximal region of the mouse gastrointestinal tract because at least some strains can adhere to and colonise the surface of the stratified squamous epithelium lining the oesophagus and forestomach. Colonisation of these surfaces results in a layer of *Lactobacillus* cells associated with the epithelium (Schaedler *et al.* 1965; Savage *et al.* 1968; Lee *et al.* 1971). Lactobacilli shed from this layer inoculate the digesta as it flows past so that these bacteria are present throughout the length of the gastrointestinal tract. The *Lactobacillus* population of the murine gastrointestinal tract remains at a high level throughout the remainder of the animal's life. Enterococci and *E. coli* are confined to the distal regions of the intestinal tract, probably because they cannot adhere to the epithelium of the proximal tract and are flushed through the small intestine at a rate that does not permit the buildup of appreciable numbers of bacteria. Although confined to the intestinal lumen in adult animals, these facultatively anaerobic bacteria form microcolonies in the mucus covering the epithelium of the distal intestinal tract of neonates (Savage *et al.* 1968). Microaerophilic, *Campylobacter*-like, spiral-shaped bacteria also colonise the mucus of the distal intestinal tract and persist in adulthood (Davis *et al.* 1983). Obligately anaerobic bacteria do not establish in the gastrointestinal tract until the infant begins to supplement its milk diet by nibbling on solid food (an allogenic factor influencing colonisation). *Bacteroides* species and fusiform-shaped bacteria (clostridia, fusobacteria) colonise the large intestine at this stage, many types associating with the mucus layer covering the intestinal surface. The obligate anaerobes attain populations of $10^{10}$ per gram of large intestine and become, therefore, the numerically dominant microbes in this region of the tract. The establishment of bacteroides and fusiforms results in a marked reduction in enterococcal and *E. coli* numbers due to the production of short chain volatile fatty acids (particularly butyric acid) by the obligate anaerobes that are inhibitory to the facultative anaerobes (an autogenic factor influencing colonisation; Lee and Gemmel 1972). Populations of these latter bacteria decrease to about $10^4$ per gram of large intestine and remain at this lower level thereafter. The microcolonies of enterococci and *E. coli* in the mucus layer are presumably obliterated by the mass of fusiforms that form a mucus-associated layer in the large intestine. Filamentous, segmented bacteria that attach by one end to the villous surface through the formation of an invagination of the enterocyte membrane colonise the ileum of mice after weaning which usually occurs at the end of the third week after birth (Davis and Savage 1974). In some colonies of mice, a yeast (*Candida pintolopesii*) establishes in the stomach after weaning and associates with the epithelial surface of the secretory mucosa (Savage and Dubos 1967). By the end of the fourth week after birth, therefore, the biological succession is complete and a microflora characteristic of an adult mouse is present in the gastrointestinal tract. This microflora can be detected in adult mice throughout the remainder of their life as long as they are maintained free of stress under standard conditions (reviewed by Tannock 1983).

## Acquisition of the Gastrointestinal Microflora in other Animals

The colonisation of the gastrointestinal tract by microbes follows the same general pattern to that observed with mice in the other animal species that have been studied (Smith and Crabb 1961). *Escherichia coli* and enterococci are numerous during early life but decrease as other microbial types become established. The adult microflora of different animal species differs in composition, however, possibly because of host dietary and physiological differences, although obligately anaerobic bacteria are always the numerically predominant microbes inhabiting the large intestine (Smith 1965). The distribution of the microflora in the gastrointestinal tract is markedly different between animal species, mostly reflecting anatomical differences between hosts. Ruminants, for example, harbour an extensive microflora in the proximal part (rumen) of the gastrointestinal tract as well as in the large intestine, whereas the human stomach and small intestine contain only low numbers of microbes. Fowl, pigs, mice and rats have numerous lactobacilli in the proximal region of the gastrointestinal tract because of the presence of stratifed, squamous epithelia to which they can adhere (reviewed by Savage 1977; reviewed by Hobson and Wallace 1982; reviewed by Tannock 1990). The climax community thus reflects the selection of specific microbial strains from a heterogeneous collection that initially invades the gastrointestinal lumen of the neonate. The selected strains possess attributes that enable them to thrive under the conditions existing in a particular type of gastrointestinal tract. Adaptation of microbes to life in association with the gastrointestinal tract of a particular animal species is well exemplified by lactobacilli: strains that associate with epithelia of fowl or of pigs will not colonise rodent epithelia and conversely. Even more host specific, filamentous segmented microbes from mice will not colonise the intestinal tract of rats and vice versa (Tannock *et al.* 1982; Tannock *et al.* 1984). Autogenic factors operating within habitats in the gastrointestinal tract are doubtless of major importance in the outcome of the colonisation process yet, apart from the clearly demonstrated involvement of short chain fatty acids under appropriate conditions of pH and $E_h$ in amensalism in the large intestine, specific in vivo examples are unavailable (reviewed by Tannock 1981; Tannock 1984). Microbial interactions in natural environments are difficult to investigate scientifically, and the results of in vitro experiments are notoriously unreliable indicators of microbial activities in natural habitats (Hentges and Freter 1962).

## Succession within a Succession

The acquisition of the normal microflora of the gastrointestinal tract is usually described in terms of qualitative and quantitative changes at the taxonomic level of bacterial genera. Differentiation between species of a single bacterial genus, or between strains of a single species, however, reveal a more complex situation. According to Mitsuoka (1989), in Japan, *Bifidobacterium breve* and *Bifidobacterium infantis* are the numerically dominant bifidobacteria in the intestinal tract of human infants, but *Bifidobacterium longum* and *Bifidobacterium adolescentis* are dominant in the case of adults. Enterococci inhabiting the gastrointestinal tract of mice belong to two species: *Enterococcus faecium* and *Enterococcus faecalis*. In infant mice, all of the *E. faecium* isolates ferment xylose whereas 43 per cent of isolates from adults do not (Tannock 1979). Plasmid profiling, a technique that involves the demonstration of the plasmid molecules characteristic of a particular bacterial isolate, enables different strains of lactobacilli to be recognised. Tannock and colleagues (1990a) used this technique to examine the colonisation of the neonatal piglet gastrointestinal tract by lactobacilli and observed a complex sequence of events as piglets of a single litter matured. The pars oesophageal epithelium in the stomach was colonised by lactobacilli within 24 h of birth. The collection of *Lactobacillus* strains, as revealed by plasmid profiling, was different in the two one-day-old piglets that were examined. At four days of age, piglets harboured different plasmid profile types of lactobacilli to those detected at 24 h, but a similar collection of strains was present in both animals examined. Yet different *Lactobacillus* strains belonging to the species *Lactobacillus acidophilus* and *Lactobacillus fermentum* had colonised the pars oesophagea by seven days after birth and one particular plasmid profile type of *Lactobacillus acidophilus* had become dominant in the two animals examined. This same strain was numerically dominant in piglets examined 14 days after birth. Even when a bacterial genus or species appears to be stably colonising the gastrointestinal tract, therefore, changes in the composition of the strains that make up the total population can be occurring. Stability in the composition of bacterial populations may be achieved eventually in habitats, but several studies involving human subjects suggest that *E. coli* populations continue to change in strain composition even in adults (reviewed by Mason and Richardson 1981). While stability in species composition of the normal microflora may be common, stability at the level of bacterial strains may be less common. If this is the case, the acquisition of the normal microflora may be never ending as new strains of endogenous or exogenous origin proliferate and attain dominance in the gastrointestinal tract under the influence of particular allogenic or autogenic factors. This is a topic that requires further research.

## Acquisition of the Human Intestinal Microflora: a Controversial Topic

The acquisition of the intestinal microflora by human infants born in developed countries has been a controversial topic in recent years. This controversy relates to the colonisation of the intestinal tract of infants consuming cow's milk-based formula feeds rather than being suckled at the breast. Conflicting results relating to the populations of bifidobacteria and *E. coli* in the faeces of breast milk and formula fed infants have been published, with some studies interpreted to show that large populations of bifidobacteria only occur in breast milk-fed babies and that *E. coli* populations are lower in these infants compared to those that received a formula feed (reviewed by Cooperstock and Zedd 1983).

The "classical" studies concerning the infant intestinal microflora are generally considered to be those of Tissier (1905). He divided the colonisation of the intestinal tract of suckled infants into three phases. In the first phase which consisted of the first few hours of life, the faeces were devoid of microbes. The second phase began between the tenth and twentieth hour of life with the detection of a heterogeneous collection of microbial types in the faeces. After three days, by which time milk had passed through the length of the intestinal tract, the third colonisation phase began. A gram-positive bacillus became the numerically dominant microbe in the faeces at this stage as judged by microscope examination of faecal smears. The other microbial types disappeared in a fairly constant order, and from the start of the third or fourth day of life until weaning, the collection of microbes in the faeces remained the same. According to Tissier, the collection seemed "to be constituted, by microscopic examination of only one species, *Bac. bifidus*, a strictly anaerobic bacterium. It is necessary to do a complete bacteriological examination to see that there exists, besides this species, other facultatively anaerobic bacteria in very limited numbers: *Bact. coli* (variété *commune*), *Enterocoque* and sometimes *Bact. lactis aerogenes*." The faeces of infants not suckled at the breast, in contrast, continued to contain a mixture of microbial types in which no single type predominated, even after the fourth day of life. To quote Tissier, "this usual flora is constituted like this: besides *Bac. bifidus*, *Bact. coli* (v. *commune*), *Enterocoque*, there exists in equal numbers *Bac. acidophilus*, *Bac. exilis*, more rarely *Staphylocoque blanc*, *Sarcines*, *Bact. lactis aerogenes*, *Bact. coli* (v. *communior*), *Levure blanche* and sometimes *Bac. III de Rodella* and *B. coli* (v. *typhimorphe*)."

**Table 1.1.** Comparison of the faecal microflora of infants of different ages consuming human milk or formula feed. Data from Stark and Lee 1982

| Bacterial group | | Age (weeks) 4 | 14 | 20 | 37 | 52 |
|---|---|---|---|---|---|---|
| Bifidobacteria | BF* | 10.6** | 10.1 | 10.2 | 10.1 | 9.9 |
| | FF | 10.3 | 10.4 | 9.8 | 9.9 | 10.4 |
| Bacteroides | BF | <3.0 | <3.0 | 9.3 | 8.2 | 9.9 |
| | FF | 9.3 | 8.4 | 8.7 | 9.6 | 9.7 |
| Anaerobic gram- | BF | <7.0 | <7.0 | <7.0 | 8.2 | 10.1 |
| positive cocci | FF | <7.0 | <7.0 | 9.1 | 10.1 | 9.9 |
| Clostridia | BF | <3.0 | <3.0 | 5.2 | 5.1 | 5.2 |
| | FF | 6.4 | 5.3 | 4.7 | 5.6 | 4.9 |
| Enterobacteria | BF | 6.1 | 5.5 | 8.1 | 8.6 | 6.9 |
| | FF | 9.4 | 8.3 | 8.5 | 8.3 | 7.6 |
| Enterococci | BF | 6.3 | 6.5 | 8.5 | 8.6 | 7.4 |
| | FF | 9.6 | 9.2 | 9.5 | 7.9 | 7.5 |

* BF = Breast milk; FF = Formula

** Mean $\log_{10}$ viable count per gram (wet weight) of faeces from infants harbouring the bacterial group

Tissier's turn-of-the century microscope observations of the faeces of breast milk fed infants hold true today, but the situation regarding formula fed babies appears to have altered. Well documented modern studies show that bifidobacteria are just as likely to be numerically dominant in the faeces of formula-fed as in breast-fed infants (Tables 1.1, 1.2). There is considerable infant to infant variation in the population size of particular bacterial genera during the first week of life in both infant groups (Tables 1.3 and 1.4) which may have contributed to the somewhat variable interpretations of the status of the infant microflora reported in the literature. More consistent values are obtained in babies older than one week, however, and realistic comparisons between infant groups are possible. The critical test as to whether biologically significant differences exist between the microfloras of breast milk or formula fed babies is whether it is possible to predict, on the basis of the microbiological study of the faeces, what the diet of the infant consisted of. Identification of the diet is not possible using data concerning the presence or absence of particular bacterial groups in infant faeces (Tables 1.2, 1.5).

**Table 1.2.** Comparison of faecal microflora of infants aged four to seven weeks consuming breast milk or formula feed. Data from Benno *et al.* 1984

| Bacterial group | | Mean $\log_{10}$ count per gram of faeces in which bacterial group was detectable | Incidence in faeces (out of 35 in each group) |
|---|---|---|---|
| Bifidobacteria | BF* | 10.7 | 35 |
| | FF | 10.6 | 33 |
| Bacteroides | BF | 8.9 | 19 |
| | FF | 9.9 | 23 |
| Peptococci | BF | 5.7 | 2 |
| | FF | 9.3 | 9 |
| Clostridia (lecithinase positive) | BF | 5.3 | 5 |
| | FF | 8.1 | 18 |
| Clostridia (others) | BF | 7.2 | 16 |
| | FF | 8.0 | 33 |
| Enterobacteria | BF | 8.2 | 34 |
| | FF | 9.3 | 35 |
| Streptococci | BF | 6.8 | 33 |
| | FF | 9.1 | 35 |

* BF = Breast milk; FF = Formula

**Table 1.3.** Variation in bacterial population size in the faeces of infants during the first week of life. Data from Moreau *et al.* 1986

| Bacterial group | Range of $\log_{10}$ viable counts per gram of faeces* | |
|---|---|---|
| | BF** | FF |
| Enterobacteria | 4 - 10 | 6 - 10 |
| Enterococci/streptococci | 3 - 9 | 1 - 10 |
| Bifidobacteria | 1 - 10 | 2 - 7 |
| Bacteroides | 2 - 8 | 1 - 8 |

*Values taken from graphical data
**BF = Breast milk; FF = Formula fed

**Table 1.4.** Variation in bacterial populations in infant faeces during the first week of life. Data from Stark and Lee 1982

| Bacterial group | Range of $log_{10}$ viable counts per gram of faeces | |
|---|---|---|
| | BF* | FF |
| Enterobacteria | <3.0 - 10.2 | <3.0 - 11.1 |
| Enterococci | 4.4 - 10.5 | <4.0 - 9.8 |
| Bifidobacteria | <4.0 - 12.1 | <4.0 - 10.4 |
| Bacteroides | <3.0 - 11.2 | <3.0 - 7.3 |

*BF = Breast milk; FF = Formula fed

**Table 1.5.** Incidence at detectable levels of bacterial groups in faeces of breast milk or formula fed infants of different ages. Data from Stark and Lee 1982

| Bacterial group | | Age (weeks) | | | | | |
|---|---|---|---|---|---|---|---|
| | | 1 | 4 | 14 | 20 | 37 | 52 |
| Bifidobacteria | BF* | 6/7** | 6/6 | 7/7 | 7/7 | 5/5 | 7/7 |
| | FF | 5/7 | 4/6 | 6/6 | 5/7 | 7/7 | 7/7 |
| Bacteroides | BF | 4/7 | 1/6 | 1/7 | 3/7 | 5/5 | 7/7 |
| | FF | 1/7 | 2/6 | 3/6 | 4/7 | 6/7 | 7/7 |
| Anaerobic gram- | BF | 1/7 | 0/6 | 1/7 | 1/7 | 3/5 | 6/7 |
| positive cocci | FF | 0/7 | 1/6 | 0/6 | 2/7 | 4/7 | 5/7 |
| Clostridia | BF | 6/7 | 0/6 | 0/7 | 3/7 | 5/5 | 7/7 |
| | FF | 4/7 | 4/6 | 5/6 | 6/7 | 6/7 | 7/7 |
| Enterobacteria | BF | 6/7 | 6/6 | 7/7 | 7/7 | 5/5 | 7/7 |
| | FF | 6/7 | 6/6 | 6/6 | 7/7 | 7/7 | 7/7 |
| Enterococci | BF | 7/7 | 6/6 | 7/7 | 7/7 | 5/5 | 7/7 |
| | FF | 6/7 | 6/6 | 7/7 | 7/7 | 7/7 | 7/7 |

* BF = Breast milk; FF = Formula

** Number of infants in which bacterial group was detected/number of infants examined

**Table 1.6.** Comparison of selected bacterial populations present in the faeces of infants aged approximately four weeks. Data from Stark and Lee 1982 and Benno *et al.* 1984

| Bacterial group | Range of $log_{10}$ viable counts per gram of faeces | |
|---|---|---|
| | BF* | FF |
| Clostridia (Stark and Lee) | <3.0 - <3.0 | <3.0 - 7.1 |
| Clostridia (lecithinase positive, Benno *et al*) | 5.3 ± 1.5** | 8.1 ± 1.2 |
| Enterobacteria (Stark and Lee) | 3.3 - 8.3 | 8.6 - 10.2 |
| Enterobacteria (Benno *et al*) | 8.2 ± 1.1 | 9.3 ± 0.7 |
| Enterococci (Stark and Lee) | 5.4 - 8.3 | 9.1 - 10.2 |
| Enterococci/Streptococci (Benno *et al*) | 6.8 ± 1.5 | 9.1 ± 0.7 |

* BF = Breast milk; FF = Formula fed

** Mean and standard deviation

Population sizes of certain bacterial groups do appear to be characteristic of the infants' diet when the data of Stark and Lee (1982)and of Benno *et al* (1984) are compared (Table 1.6). For infants about four weeks of age, clostridial counts greater than $10^7$, and enterococcal/streptococcal and enterobacterial counts of greater than $10^9$ per gram of faeces are likely to have originated from formula fed babies. Examination of data from other studies where only one or a few bacterial groups were enumerated and in which the age of the individual babies was given, shows that this prediction holds true for the clostridia (Bullen and Willis, 1971; Bullen *et al.* 1976; Bullen *et al.* 1977; Mitsuoka 1982; Simhon *et al.* 1982; Kawase *et al*, 1983; Lundequist *et al*, 1985) but not for the enterobacteria or enterococci whose numbers are extremely variable in both breast and formula fed infants. The clostridial population, however, is always lower in breast fed, compared to formula fed, infants. The significance of this difference to the infants is not known. Clostridia may be involved in the causation of necrotising enterocolitis, but the peak incidence of this disease is in infants of four days, rather than four weeks, of age (Pedersen *et al.* 1976; Kindley *et al.* 1977).

It is not appropriate to apply Tissier's observations on the faecal microflora of infants not suckled at the breast to the modern situation. Improvements in bacteriological culture techniques and in the nutrition of bottle-fed infants make such comparisons unreasonable. Regarding infant nutrition, most babies during the nineteenth century were suckled at the mother's breast although the inferior quality and quantity of milk often available to the child contributed to the high infant mortality of that period. Some children were suckled by wet nurses, but availability of suitable and reliable nurses was a problem. The only alternative for mothers who could not breast feed their child was to prepare paps (flour or breadcrumbs cooked in water or milk), panada (bread, broth or milk, sometimes with eggs added) or cow's milk diluted with water and sweetened with sugar. The quality of cow's milk was extremely variable, especially in cities, depending on the health of the cows, the honesty of the dairy owner (watered milk) and the lack of hygiene in the dairy and whether storage conditions for the milk were adequate. From the 1860's, infant foods were marketed, usually based on the product devised by the German chemist, Justus von Liebig. This food contained wheat flour, cow's milk, malt flour and bicarbonate. During the cooking process, the malt converted the starch flour to dextrin and glucose which could be more easily assimilated by the infant intestinal tract (Fildes 1986; Apple 1987). This was the type of infant food available in Tissier's time. Condensed and evaporated milks were possible alternatives. The aim of modern research in formula feeds has been to manufacture products that duplicate the properties of human milk. The formulas have energy values that are similar to the currently accepted average for human milk, about 50% of the energy being derived from fat. Butterfat from cow's milk is removed during manufacture and replaced by blends of animal and vegetable fats so that the fatty acid composition and ratios of saturated to unsaturated fatty acids of the formula is nearer that of human milk. Linoleic and alpha linoleic acids provide at least one percent of the total energy since they have been shown to be essential fatty acids for human infants. Up to 12% of energy is derived from protein. Specific amino acids are added to approximate their quantities found in human milk and to ensure that essential amino acids are provided. Lactose, sucrose and maltodextrins are added to raise the carbohydrate level to that of human milk. Formulas are fortified by the addition of vitamin D and iron (Anon. 1988). Thus modern studies of the acquisition of the gastrointestinal microflora of human infants receiving formula feeds should not be expected to be comparable to those of Tissier.

The incidence of gastrointestinal upset is said to be lower in breast fed infants than in those consuming formula feeds (Bullen and Willis 1971). It is by no means clear whether this observation holds true in the 1990's, but if so, this difference may be due to the different physicochemical properties of human and cow's milk (Bullen and Willis 1971). The overall effect of these different properties appears to be a reduced buffering capacity of the faeces of formula fed infants. The pH of the faeces from breast milk fed infants tend to be lower, more than 30% of such infants having a faecal pH of less than 5 (Bullen and Willis 1971). The addition of oligosaccharides (TOS: transgalactosylated oligosaccharides) to formulas is aimed at achieving a lowering of the faecal pH since these carbohydrates pass through the small intestine without alteration but are fermented by bacteria in the large intestine (Tanaka et al. 1983). Human milk contains a large number of antibacterial, antiviral and antiparasite factors (e.g. antibodies and complement, lactoferrin, lactoperoxidase, lysozyme, macrophages, neutrophils and lymphocytes, haemagglutins, lipids). These factors are absent from, or are at low levels in, formula feeds (May 1988; Raibaud 1988). Human milk also contains epidermal growth factor, nerve growth factor, somatomedin-C, insulin-like growth factor, insulin, thyroxine, cortisol, taurine, glutamine, and amino sugars, all of which are believed to promote the maturation of the gastrointestinal tract in neonates. Lack of exposure of the intestinal mucosa to these molecules may delay maturation of the infant gastrointestinal tract rendering it more susceptible to perturbations (Sheard and Walker 1988).

## Acquisition of the Microflora Influences Host Biochemistry

The biochemistry of the gastrointestinal tract is markedly influenced by the metabolic activities of the normal microflora. The digestion of plant structural molecules by microbes inhabiting the rumen has been well studied and described and it is clear that the short chain volatile fatty acids produced by rumen microbes provide the major source of energy for the ruminant host (Hobson and Wallace 1982). Comparison of germfree monogastric hosts with their conventional counterparts has demonstrated that the microbes in the gastrointestinal tract also contribute to the biochemical characteristics of the animal host. The normal microflora influences the enzymatic activity of the intestinal contents (e.g. ß-D-glucuronidase, azoreductase, bile salt hydrolase), short chain volatile fatty acid concentration, oxidation-reduction potential, host physiology (e.g. rate of replacement of enterocytes), immunology (e.g. stimulation of reticuloendothelial tissues), and the modification of host-synthesised molecules (e.g. reduction of cholesterol to coprostanol, dehydroxylation of primary bile acids to form secondary bile acids, mucin degradation). Thus the acquisition of the normal microflora can be followed by measuring biochemical parameters as well as by detecting and enumerating particular groups of microbes in the laboratory (Gordon and Pesti 1971; Tannock *et al.* 1989).

Germfree rodents have a high level of proteolytic activity in large intestinal contents compared to conventional animals. This is because, in the germfree animal, trypsin secreted in the pancreatic juice passes down the small intestine and remains active in the large bowel. In conventional rodents, the tryptic activity is reduced because the members of the normal microflora inactivate the proteolytic enzyme (Midtvedt 1985). Inactivation of tryptic activity in the caecum of rats occurs as early as ten days after birth, indicating that an appropriately complex microflora required for this function establishes early in the biological succession. Faeces collected from adult humans do not generally have tryptic activity. Human infants, however, usually have faecal tryptic activity which can be regularly detected until the children are about 20 months of age. A decrease or absence of tryptic activity has been observed in the majority of children sampled between 46 and 61 months of age which suggests that the development of the trypsin-inactivating microflora in humans is prolonged (Norin *et al.* 1986). Microbes capable of reducing cholesterol to coprostanol do not colonise the intestinal tract of humans until after the children are one year old. Bilirubin-metabolising and mucin-degrading microbes colonise the intestinal tract of the majority of infants before one year after birth (Norin *et al.* 1985). Mucin modification begins in the mouse colon from the time at which obligately anaerobic bacteria become established (Hill *et al.* 1990). Overall, these observations suggest that it is the obligately anaerobic inhabitants of the intestinal tract that are responsible for much of the modifications to host-derived molecules that occur in the intestinal habitat.

The type of short chain volatile fatty acids detected in infant faeces changes according to the type of diet consumed. Acetic and propionic acid are commonly detected in the faeces of breast milk fed infants, but isobutyric, butyric, isovaleric, valeric and isocaproic acids are present only after supplementary feeding begins, reflecting changes in the types of bacteria colonising the intestinal tract as the biological succession proceeds (Bullen *et al.* 1976). Supplemental feeding of breast milk fed babies also appears to initiate a decrease in oxidation-reduction potential of the faeces. The Eh of meconium is about +175 mV, but the faeces from infants one to two days old are more reduced (-113 mV). Adult values (-348 mV) are not reached until after weaning (reviewed by Cooperstock and Zedd 1983).

# The Application of Knowledge of the Acquisition of the Normal Microflora

Studies of the acquisition of the microflora are not only exercises in descriptive science, important though it is to record and understand natural phenomena. The data obtained in such studies is of practical significance in the development of probiotic products, in understanding why neonates are more susceptible to infection by certain pathogens compared to adults, and in investigating the epidemiology of plasmids that encode antibiotic resistance in bacteria and hence in the design of regimes for the rational use of antibiotics.

## The Development of Probiotics

The derivation of preparations of living microbes that are to be fed to farm animals or human beings with the aim of stimulating productivity and health appears to rely on the concept that the probiotic will be administered on a daily basis to the recipients (Fuller 1989). Persistence of probiotic microbes in the gastrointestinal tract of adult animals requires continual administration of the product because of the phenomenon of microbial interference in which microbes already established in an ecosystem prevent the colonisation of that site by newly introduced organisms (reviewed by Tannock 1981; Tannock 1984). Thus the normal microflora of the adult intestinal tract confers a degree of resistance to infection by *Salmonella* species and other pathogens. The mechanisms that regulate the intestinal microflora prevent the establishment of these allochthonous microbes: probiotic microbes suffer the same fate as small inocula of *Salmonella* or other intestinal pathogens.

Probiotic microbes could possibly colonise the gastrointestinal tract (i.e. occupy an ecological niche) if they were introduced into the germfree habitats of the neonatal animal and possessed appropriate colonisation characteristics. In simple terms, the first qualified microbe to enter the habitat and to thrive would be numerically dominant thereafter. This approach appears to be successful in the case of *E. coli* since Duval-Iflah and colleagues (1982) have reported that the colonisation of the intestinal tract of human infants by a plasmid-free strain reduced the likelihood of antibiotic resistant enterobacteria colonising the babies. This concept does not hold true for lactobacilli, however, since experiments with piglets have demonstrated that strains of intestinal lactobacilli used as inocula for neonatal piglets are numerically dominant in the gastrointestinal ecosystem for only one day after a single administration and persist for only a limited time (Pedersen and Tannock 1989). The failure of introduced lactobacilli to persist can be explained once it is understood that a succession of *Lactobacillus* strains colonises the piglet gastrointestinal tract after birth (Tannock *et al.* 1990a). Different strains predominate as the piglet matures, perhaps reflecting physiological changes in the animal, or changes in the composition of the sow's milk. It is clear that the factors that enable lactobacilli to colonise piglets of different ages must be elucidated if reliable probiotics for use with these animals are to be developed.

A knowledge of the normal microflora of neonatal animals is also necessary for the development of probiotics to be used to inoculate young animals whose acquisition of a microflora may be delayed or prevented. Chickens raised in modern hatcheries never come into contact with an adult fowl and therefore do not acquire an intestinal microflora as rapidly as birds in a farmyard (Fuller 1989). Inoculation of chickens with an appropriate intestinal microflora can increase the resistance of the birds to infection by *Salmonella* species (Schneitz *et al.* 1990). Human infants delivered by sterile caesarian section, or premature babies placed in incubators postnatally may similarly be deprived of a microflora during the first days of life. It has been proposed that such infants would benefit by being inoculated *per os* with an appropriate microflora (Bennet and Nord, 1987; Hall *et al.* 1990).

## Biological Freudianism

"Biological Freudianism" was a term used by Dubos and colleagues to denote the lasting effects of early environmental influences (Dubos *et al.* 1966). The quality of nutrition received prenatally and during the first years of life markedly affects growth rate, age at which sexual maturity is attained, and resistance to infectious diseases. Susceptibility to infection by intestinal pathogens is high in malnourished children in developing countries. In Bangladesh, for example, the incidence of diarrhoea caused by enterotoxigenic *E. coli* ranges from one to two episodes per child per year during the first five years of life (Black *et al.* 1982). These severe infections are amongst the main causes of death in children in developing countries. The infections are also a major determinant of growth retardation and malnutrition in survivors because, due to intestinal epithelial damage causing malabsorption of nutrients, they aggravate an already malnourished state. These effects remain evident throughout the life of the individual. Of pertinence in relation to the acquisition of the normal microflora, are the observations made by Schaedler and Dubos (1962) relating to a lasting influence exerted by the composition of the normal microflora of mice. During the 1960's, a specific-pathogen-free colony of Swiss mice (NCS mice) was derived at the Rockefeller Institute in New York. The mice were maintained under conditions of high sanitation and, unlike the mice from which they were derived (SS mice), did not harbour *E. coli, Proteus vulgaris*, or *Pseudomonas* species as members of the gastrointestinal microflora. The two types of mice were found to differ markedly in their response to the parenteral administration of endotoxin. NCS mice were tolerant to doses that were lethal when administered to SS mice. Exposure to SS mice during early life, or prior 'vaccination' with heat-killed gram-negative bacteria (providing small doses of endotoxin), increased the susceptibility of the NCS mice so that it equaled that of the SS animals. Thus a microflora containing *E. coli, Proteus* and *Pseudomonas* (bacterial types that are numerically dominant in the murine intestinal tract early in life) sensitised the animals so that as adults they were highly susceptible to endotoxin.

Comparison of the immunological characteristics of germfree and conventional animals has shown that the presence of the normal microflora stimulates the reticuloendothelial tissues of the host, particularly by those tissues associated with the gastrointestinal tract. Peyer's patches and mesenteric lymph nodes are more developed, the lamina propria of the intestinal mucosa contains more neutrophils and lymphocytes, and the serum contains a higher concentration of gamma globulin in conventional animals compared to germfree (reviewed by Gordon and Pesti 1971). The role of specific microbes in stimulating the immunological tissues of the animals during the acquisition of the normal microflora is not known but the cumulative effect of the exposure of gastrointestinal tissues to microbial antigens and the translocation of antigenic material and small numbers of bacterial cells from the intestinal lumen is marked. Probably mainly through the activation of macrophages by bacterial antigens, the conventional animal is said to be primed immunologically and better able to respond rapidly to the presence of an invading pathogenic microbe. Such priming of resistance mechanisms is of lifelong significance to the animal host.

## Colonisation by Potential Pathogens

The microflora, although considered to be normal in the sense that the microbes colonise clinically healthy subjects, contains species that are opportunistic pathogens. Knowledge of the acquisition of the normal microflora, together with the fact that the immunological attributes of neonates are immature and lack prior exposure to specific pathogens, provides an explanation for the occurrence of certain infections in neonatal animals. Acute purulent meningitis in human infants less than one month of age is commonly due to *Escherichia coli* (Ray 1990). The prevalence of this bacterial species in these infections is understandable when it is considered that infants can be exposed to *E. coli* during passage through the vagina, and that they harbour large populations of *E. coli* in their alimentary tract during the first weeks of life. The tissues of the infant are therefore exposed to large numbers of potentially pathogenic cells. Similarly, intestinal infections due to enterotoxigenic *E. coli* have a high incidence in neonatal farm animals. Mechanisms that maintain enterobacterial populations at low levels are not yet operating in the neonatal intestinal tract. Thus large populations of *E. coli* are present in the faeces of the animals, ensuring ease of transmission of infection between individuals if a pathogenic strain of *E. coli* should be present. *Candida albicans* is a member of the normal microflora of the human oral cavity that often causes acute pseudomembranous candidiasis (oral thrush) in infants. In these cases, bacterial members of the microflora that normally suppress the replication of the yeast are lacking and, coupled with the immunological immaturity of the infant, *Candida albicans* is able to proliferate and invades the oral mucosa (Thea and Keusch 1989).

In contrast to the involvement of certain members of the normal microflora in the production of disease in neonates, but rarely in adults unless predisposing conditions to infection are present (e.g. immunosuppression, prolonged antibiotic therapy administered orally), human infants can harbour a moderately large population ($10^5$ to $10^6$ per gram) of *Clostridium difficile* in the intestinal tract without apparent harm (reviewed by Bartlett 1983). Populations of this size in the adult intestinal tract are associated with pseudomembranous colitis in which severe diarrhoea is produced through the production of toxins by the clostridium. The reason for the resistance of infants to the clostridial toxins is not known. Since the toxins are detectable in the infant intestinal tract it appears that receptors for them may not be present in the neonatal intestinal epithelium.

## Antibiotic Resistance Transfer in the Gastrointestinal Tract

Although antibiotic resistance determinants (r-determinants) were present in bacterial populations prior to the widespread use of antimicrobial drugs in human and veterinary medicine, epidemiological studies have shown that the dissemination of r-determinants has become widespread in bacterial populations since the 1950's (Hardy 1986). Molecular analysis of r-determinants detected in different bacterial hosts of clinical importance has demonstrated that interspecies and intergeneric transfer of DNA encoding antibiotic resistance has occurred in natural ecosystems (Levy 1986). The study of the conditions under which r-determinants are transferred between bacterial types in habitats is therefore important if methodologies for the rational use of antibiotics in the treatment and prophylaxis of infectious diseases are to be devised. The gastrointestinal tract is inhabited by several hundred species of bacteria, potentially any of which could act as a reservoir of r-determinants that might be passed to pathogens. Infections caused by antibiotic resistant pathogens are more difficult, and more costly, to treat because alternative antibiotics to those commonly used in the clinical situation must be resorted to. The use of antibiotics in the treatment of infectious diseases and as prophylactic agents in human and veterinary medicine, and as 'growth promoting' drugs in animal husbandry, provides sources of antibiotics that may influence the derivation and transfer of transmissable DNA molecules in the gastrointestinal tract (Linton 1986). McConnell and colleagues (1991) recently followed the fate of the broad-host-range, conjugative plasmid pAMß1 in the digestive tract of mice by DNA-DNA hybridisations using a probe containing the *ermAM* determinant (macrolide-lincosamide-streptogramin type B resistance). Transfer of pAMß1 from *Lactobacillus reuteri* to *Enterococcus faecium* occurred in the digestive tract of infant mice, but not in that of adults. The intestinal microflora of infant mice differs qualitatively and quantitatively from that of adult animals. Lactobacilli are numerous in the digestive tract of both adult and infant mice, but enterococci only achieve a large population in infant mice. Transfer of plasmid pAMß1 ocurred between the lactobacilli and enterococci in the digestive tract from as early as eight days after birth. In vitro experiments revealed that the efficiency of transfer of pAMß1 between the bacterial strains was low (about $10^{-8}$). The transfer occurred in the infant gastrointestinal tract, therefore, because only in these animals had the enterococci achieved a population size suitable for conjugation events to occur between potential donor (lactobacilli) and potential recipient (enterococci). An explanation of why plasmid transfer occurred only in the infant digestive tract was possible in this instance because the difference between infant and adult microfloras was known.

## Conclusions

i The acquisition of the normal microflora of all animal species that have been studied follows a similar pattern. Gram-positive lactic acid-producing bacteria and facultatively anaerobic bacteria are numerous in the large intestine while the infant is fed a diet composed exclusively of maternal milk. Supplemental feeding marks the beginning of the colonisation of the intestinal tract by obligate anaerobes that profoundly influence the biochemistry of the intestinal milieu. After weaning, a microflora characteristic of an adult animal of the species gradually develops.

ii A knowledge of the gastrointestinal microflora of neonatal animals is important for the development of probiotics, in understanding the susceptibility of neonates to particular infections, in evaluating the use of antibiotics in human and veterinary medicine, and in appreciating the intimate and complex relationships that are established between the animal host and its normal microflora from the moment of birth.

## References

Anon (1988) Present day practice in infant feeding: third report. Report on Health and Social Subjects, Department of Health and Social Security, Her Majesty's Stationary Office, London

Apple RD (1987) Mothers and Medicine. University of Wisconsin Press, Madison

Bartlett JG (1983) *Pseudomembranous colitis*. In: Hentges DJ (ed) Human Intestinal Microflora in Health and Disease, Academic Press, New York pp 447-479

Bennet R, Nord CE (1987) Development of the faecal anaerobic microflora after caesarean section and treatment with antibiotics in newborn infants. Infection 15:332-336

Benno Y, Sawada K, Mitsuoka T (1984) The intestinal microflora of infants: composition of fecal flora in breast-fed and bottle-fed infants. Microbiol Immunol 28:975-986

Black RE, Brown KH, Becker S, Alim ARMA, Huq I (1982) Longitudinal studies of infectious diseases and physical growth of children in rural Bangladesh. Am J Epidemiol 115:315-324

Brunel A, Gouet P (1982) Kinetics of the establishment of gastrointestinal microflora in the conventional rat. Ann Microbiol (Inst Pasteur) 133:325-334

Brunel A, Gouet P (1989) Intestinal microflora of the newborn rat as related to mammary, faecal, and vaginal staphylococci strains isolated from the dam. Can J Microbiol 35:989-993

Bullen CL, Willis AT (1971) Resistance of the breast-fed infant to gastroenteritis. Brit Med J 3:338-343

Bullen CL, Tearle PV, Willis AT (1976) Bifidobacteria in the intestinal tract of infants: an in-vivo study. J Med Microbiol 9:325-333

Bullen CL, Tearle PV, Stewart MG (1977) The effect of "humanised" milks and supplemented breast feeding on the faecal flora of infants. J Med Microbiol 10:403-413

Carlsson J, Gothefors L (1975) Transmission of *Lactobacillus jensenii* and *Lactobacillus acidophilus* from mother to child at time of delivery. J Clin Microbiol 1:124-128

Cooperstock MS, Zedd AJ (1983) Intestinal flora of infants. In: Hentges DJ (ed) Human Intestinal Microflora in Health and Disease, Academic Press, New York pp 79-99

Davis CP, Savage DC (1974) Habitat, succession, attachment, and morphology of segmented, filamentous microbes indigenous to the murine gastrointestinal tract. Infect Immun 10:948-956

Davis CP, McAllister JS, Savage DC (1973) Microbial colonization of the intestinal epithelium in suckling mice. Infect Immun 7:666-672

Dubos R, Savage D, Schaedler R (1966) Biological freudianism. Lasting effects of early influences. Pediatrics 38:789-800

Duval-Iflah Y, Ouriet M F, Moreau C, Daniel N, Gabilan JC, Raibaud P (1982) Implantation precoce d'une souche de *Escherichia coli* dans l'intestin de nouveau-nes humains: effet de barriere vis-a vis de souches de *E. coli* antibioresistantes. Ann Microbiol (Paris) 133A: 393-408

Fildes VA (1986) Breasts, Bottles and Babies. Edinburgh University Press, Edinburgh

Fuller R (1989) Probiotics in man and animals. J Appl Bacteriol 66:365-378

Gordon HA, Pesti L (1971) The gnotobiotic animal as a tool in the study of host microbial relationships. Bacteriol Rev 35:390-429

Hall MA, Cole CB, Smith SL, Fuller R, Rolles CJ (1990) Factors influencing the presence of faecal lactobacilli in early infancy. Arch Dis Childhood 65:185-188

Hardy K (1986) R plasmids. In: Hardy K (ed) Bacterial Plasmids, 2nd edn, American Society for Microbiology, Washington, DC pp 55-81

Hentges DJ, Freter R (1962) In vivo and in vitro antagonism of intestinal bacteria against *Shigella flexneri*. I. Correlation between various tests. J Infect Dis 110:30-37

Hill RRH, Cowley HM, Andremont A (1990) Influence of colonizing micro-flora on the mucin histochemistry of the neonatal mouse colon. Histochem J 22:102-105

Hobson PN, Wallace RJ (1982) Microbial ecology and activities in the rumen, Parts I and II. CRC Critical Rev Microbiol 9:165-225 and 253-320

Kawase K, Suzuki T, Kiyosawa I, Okonogi S, Kawashima T, Kuboyama M (1983) Effects of composition of infant formulas on the intestinal microflora of infants. Bifidobacteria Microflora 2:25-31

Kindley AD, Roberts PJ, Tulloch WH (1977) Neonatal necrotising enterocolitis. Lancet i:649

Lee A, Gemmell E (1972) Changes in the mouse intestinal microflora during weaning: role of volatile fatty acids. Infect Immun 5:1-7

Lee A, Gordon J, Lee C-J, Dubos R (1971) The mouse intestinal microflora with emphasis on the strict anaerobes. J Exp Med 133:339-352

Levy SB (1986) Ecology of antibiotic resistance determinants. In: Banbury Report 24: Antibiotic Resistance Genes: Ecology, Transfer, and Expression, Cold Spring Harbor pp 17-30

Linton AH (1986) Flow of resistance genes in the environment and from animals to man. J Antimicrob Chemother 18, suppl C:189-197

Lundequist B, Nord CE, Winberg J (1985) The composition of the faecal microflora in breastfed and bottle fed infants from birth to eight weeks. Acta Paediatr Scand 74:45-51

Mason TG, Richardson G (1981) *Escherichia coli* and the human gut: some ecological considerations. J Appl Bacteriol 51:1-16

May JT (1988) Microbial contaminants and antimicrobial properties of human milk. Microbiol Sciences 5:42-46

McConnell MA, Mercer AA, Tannock GW (1991) Transfer of plasmid pAMß1 between members of the normal microflora inhabiting the murine digestive tract and modification of the plasmid in a *Lactobacillus reuteri* host. Microb Ecol Hlth Dis 4:343-355

Midtvedt T (1985) The influence of antibiotics upon microflora-associated characteristics in man and animals. In: Wostman BS (ed) Progress in Clinical and Biological Research, vol. 181. Germfree Research: Microflora Control and its Application to the Biomedical Sciences, Alan R Liss, New York pp 241-244

Mitsuoka T (1982) Recent trends in research on intestinal flora. Bifidobacteria Microflora 1:3-24

Mitsuoka T (1989) Taxonomy and ecology of the indigenous intestinal bacteria. In: Hattori T, Ishida Y, Maruyama Y, Morita RY, Uchida A (eds) Recent Advances in Microbial Ecology, Japan Scientific Societies Press, Tokyo pp 493-498

Moreau M-C, Thomasson M, Ducluzeau R, Raibaud P (1986) Cinetique d'etablissement de la microflore digestive chez le nouveau-ne humain en fonction de la nature du lait. Reprod Nutr Develop 26:745-753

Norin KE, Gustafsson BE, Lindblad BS, Midtvedt T (1985) The establishment of some microflora associated biochemical characteristics in feces from children during the first years of life. Acta Paediatr Scand 74:207-212

Norin KE, Midtvedt T, Gustafsson BE (1986) Influence of intestinal microflora on the tryptic activity during lactation in rats. Lab Anim 20:234-237

Pedersen K, Tannock GW (1989) Colonization of the porcine gastrointestinal tract by lactobacilli. Appl Environ Microbiol 55:279-283

Pedersen PV, Hansen FH, Halveg AB, Christiansen ED, Justesen T, Hogh P (1976) Necrotising enterocolitis of the newborn - is it gas-gangrene of the bowel? Lancet ii:715-716

Raibaud P (1988) Factors controlling the bacterial colonization of the neonatal intestine. In: Hanson LA (ed) Biology of Human Milk, Raven Press, New York pp 205-219

Ray CG (1990) Central nervous system infections. In: Sherris JC (ed) Medical Microbiology, Elsevier Science Publishing, New York pp 865-872

Savage DC, Dubos RJ (1967) Localization of indigenous yeast in the murine stomach. J Bacteriol 94:1811-1816

Savage DC (1977) Microbial ecology of the gastrointestinal tract. Ann Rev Microbiol 31:107-133

Savage DC, Dubos R, Schaedler RW (1968) The gastrointestinal epithelium and its autochthonous bacterial flora. J Exp Med 127:67-76

Schaedler RW, Dubos RJ (1962) The fecal flora of various strains of mice. Its bearing on their susceptibility to endotoxin. J Exp Med 115:1149-1160

Schaedler RW, Dubos R, Costello R (1965) The development of the bacterial flora in the gastrointestinal tract of mice. J Exp Med 122:59-66

Schneitz C, Hakkinen M, Nuotio L, Nurmi E, Mead G (1990) Droplet application for protecting chicks against salmonella colonisation by competitive exclusion. Vet Rec 126:510

Sheard NF, Walker WA (1988) The role of breast milk in the development of the gastrointestinal tract. Nutr Rev 46:1-8

Simhon A, Douglas JR, Drasar BS, Soothill JF (1982) Effect of feeding on infants' faecal flora. Arch Dis Childhood 57:54-58

Smith HW (1965) Observations on the flora of the alimentary tract of animals and factors affecting its composition. J Path Bacteriol 89:95-122

Smith HW, Crabb WE (1961) The faecal bacterial flora of animals and man: its development in the young. J Path Bacteriol 82:53-66

Stark PL, Lee A (1982) The microbial ecology of the large bowel of breast-fed and formula-fed infants during the first year of life. J Med Microbiol 15:189-203

Tanaka R, Takayama H, Morotomi M, Kuroshima T, Ueyama S, Matsumoto K, Kuroda A, Mutai M (1983) Effects of administration of TOS and *Bifidobacterium breve* 4006 on the human fecal flora. Bifidobacteria Microflora 2:17-24

Tannock GW (1979) Coliforms and enterococci isolated from the intestinal tract of conventional mice. Microb Ecol 5:27-34

Tannock GW (1981) Microbial interference in the gastrointestinal tract. ASEAN J Clin Sci 2:2-34

Tannock GW (1983) Effect of dietary and environmental stress on the gastrointestinal microbiota. In: Hentges DJ (ed) Human Intestinal Microflora in Health and Disease, Academic Press, New York pp 517-539

Tannock GW (1984) Control of gastrointestinal pathogens by normal microflora. In: Klug MJ, Reddy CA (eds) Current Perspectives in Microbial Ecology, American Society for Microbiology, Washington, DC pp 374-382

Tannock GW (1990) The microecology of lactobacilli inhabiting the gastrointestinal tract. Adv Microb Ecol 11:147-171

Tannock GW, Szylit O, Duval Y, Raibaud P (1982) Colonization of tissue surfaces in the gastrointestinal tract of gnotobiotic animals by lactobacillus strains. Can J Microbiol 28:1196-1198

Tannock GW, Miller JR, Savage DC (1984) Host specificity of filamentous, segmented microorganisms adherent to the small bowel epithelium in mice and rats. Appl Environ Microbiol 47:441-442

Tannock GW, Dashkevicz MP, Feighner SD (1989) Lactobacilli and bile salt hydrolase in the murine intestinal tract. Appl Environ Microbiol 55:1848-1851

Tannock GW, Fuller R, Pedersen K (1990a) Lactobacillus succession in the piglet digestive tract demonstrated by plasmid profiling. Appl Environ Microbiol 56:1310-1316

Tannock GW, Fuller R, Smith SL, Hall MA (1990b) Plasmid profiling of members of the family *Enterobacteriaceae*, lactobacilli, and bifidobacteria to study the transmission of bacteria from mother to infant. J Clin Microbiol 28:1225-1228

Thea DM, Keusch G (1989) Digestive system. In: Schaechter M, Medoff G, Schlessinger D (eds) Mechanisms of Microbial Disease, Williams and Wilkins, Baltimore pp 628-650

Tissier H (1905) Repartition des microbes dans l'intestin du nourrisson. Ann Inst Pasteur 19:109-123

Chapter 2

# Metabolic Activities of the Normal Colonic Flora

G. T. Macfarlane and G. R. Gibson

## Introduction

The conventional view of the human large bowel as an appendage of the digestive tract, whose principal purpose was the conservation of salt and water and the disposal of waste materials, is increasingly being replaced with that of a highly specialised digestive organ, which through the activities of its constituent microbiota rivals the liver in its metabolic capacity and in the diversity of its biochemical transformations.

The last 20 years or so have been particularly exciting for those involved in gut microbiology, since they have seen significant advances in our understanding of the composition of the colonic microflora, and its symbiotic role in carbohydrate and protein metabolism, where the host provides dietary residues or endogenously produced substrates for the bacteria, which in turn supply metabolites such as short chain fatty acids (SCFA), for utilisation by the host.

## The Large Intestine

In persons living in Western countries, the main areas of permanent colonisation of the gastrointestinal tract are the terminal ileum and large intestine. This is mainly because gastric acid kills many microorganisms in the stomach, thereby reducing numbers entering the small bowel. However, rapid passage of digesta through the stomach and small bowel (approximately 4 h - 6 h) also prevents the establishment of significant microbial populations in these regions (Gorbach 1967). The rate of movement of gut contents slows markedly in the large bowel, with values in the United Kingdom ranging from 20 h to 140 h. The mean rate of transit is about 60 h (Cummings *et al.* 1993), which provides sufficient time for a complex and stable ecosystem to develop

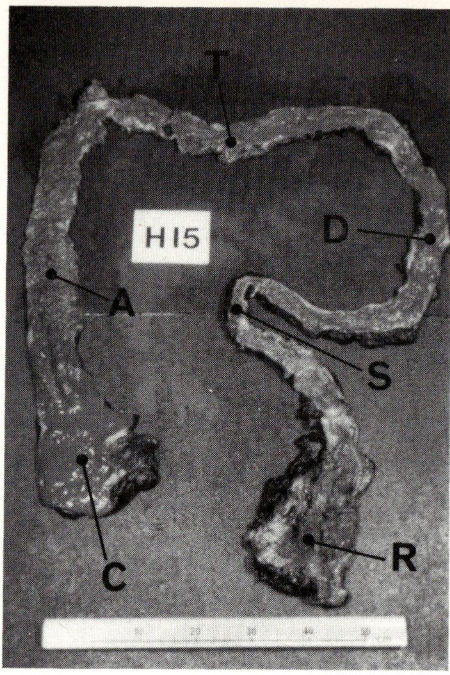


**Fig. 2.1.** The human large intestine. This was removed from a 59 year old male shortly after death. It can be seen that contents in the proximal colon have a watery consistency, whereas material in the rectum is more solid and similar to faeces. C, caecum; A, ascending colon; T, transverse colon; D, descending colon; S, sigmoid colon; R, rectum. Photograph courtesy of Dr. J.H. Cummings.

(Cummings 1978). To understand the factors that influence the growth and activities of gut bacteria it is first necessary, to consider the organ and its functions.

The human colon is typically about 150 cm long, and can be divided anatomically into the caecum, ascending colon, transverse colon, descending colon, sigmoid colon and rectum (Fig. 2.1). It has a volume of about 540 ml in the United Kingdom and receives somewhere in the region of 1.5 kg of material from the small bowel each day, most of which is water and is rapidly absorbed. The adult large gut contains around 220 g of contents, although the range (58 g - 908 g) is considerable (Banwell *et al.* 1981; Cummings *et al.* 1990; Cummings *et al.* 1993), of which about 35g is dry matter (Cummings *et al.* 1990). Due to progressive absorption of water and solutes, the moisture content of gut contents changes as they move through the different regions of the colon (see Fig. 2.1), being approximately 86% in the caecum and 77% in the sigmoid/rectum (Cummings *et al.* 1990). The average daily output of material from the colon in the United Kingdom is 120 g (Cummings *et al.* 1993) and bacteria are a major component, comprising approximately 55% of faecal solids in persons living on Western diets (Stephen and Cummings 1980), which is equivalent to 18 g of bacterial dry matter, or a total bacterial mass in the gut of 90 g. Their growth and activities are influenced to a considerable degree by the structure and physiology of the large bowel.

Material from the small intestine enters the large gut at the caecum, where there is rapid breakdown of readily digestible substrates by the bacteria. There is significant mixing of dietary residues from one day to the next, which form a pool of digesta in the proximal colon (Wiggins and Cummings 1976). Portions of this material are periodically transferred to the transverse and distal colon, where further water absorption increases the viscosity and reduces mixing. Due to utilisation of material, particularly carbohydrates, by bacteria in the proximal colon, there is a progressive reduction in substrate availability towards the distal colon, affecting such factors as the type and amounts of fermentation products that are formed and pH. This may influence a number of bacterial activities including fermentation product formation (Blackwood *et al.* 1956) and enzyme activities such as catalase, deaminases, decarboxylases, fumarase, formate, hydrogen lyase (Gale and Epps 1942), bile acid 7-α-dehydrogenase (Midvedt and Norman 1968) and some gut proteases and peptidases (Gibson and Macfarlane 1988; Macfarlane *et al.* 1988b; Allison and Macfarlane 1989a). Thus, studies on the activities of the colonic microflora should *a priori* recognise that certain metabolic processes may be restricted to, or predominate in, particular parts of the colon.

The large intestine is an open system in the sense that digesta from the small bowel enters at one end, and faeces is periodically excreted at the other. Because of this, the colon is often likened to a continuous culture system, but this is probably an oversimplification. Due to the way in which material moves through the gut, only the caecum and ascending colon can really be considered to exhibit characteristics of a continuous culture, since digesta in the remaining colon are frequently present as isolated masses (Cummings *et al.* 1987), and because there is no further input or removal of substrates or bacteria, gut contents in the distal colon are more likely to resemble batch or fed-batch systems. The proximal colon therefore effectively acts as the primary fermentation chamber and as a reservoir of bacteria.

## Composition of the Gut Microflora

Although faecal bacteria were first observed microscopically by van Leeuwenhoek in the early eighteenth century, the scale of colonisation of the large intestine by bacteria was not appreciated until quite recently. Savage (1977) has observed that about 90% of the $10^{14}$ cells associated with the human body are microorganisms, and that the vast majority of these are bacteria residing in the large intestine. Viable bacterial counts in excess of $10^{11}$ per gram are commonly found in faeces (Moore and Holdeman 1974; Finegold *et al.* 1975; Reddy *et al.* 1975), but direct microscope estimations of bacterial numbers in gut contents indicates that considerably more cells are present, and that total counts increase by an order of magnitude from the proximal to distal colon (Fig. 2.2). Possible reasons for these differences include non-viability of large numbers of cells and the inability of current cultural methods to facilitate their isolation *in vitro*.

Several hundred different species of bacteria are thought to be present in the large intestine under normal circumstances (Moore and Holdeman 1974; Finegold *et al.* 1983). The vast majority of these organisms are anaerobes, but they exhibit varying degrees of tolerance towards oxygen, ranging from relatively oxygen tolerant bacteroides (Duerden 1980) and bifidobacteria (Bezkorovainy and Miller-Catchpole 1989), to very strictly anaerobic methanogenic bacteria (Miller and Wolin 1982).

Anaerobic bacteria appear to outnumber aerobic species by a factor of about 1000. The principal anaerobic groups and some of their characteristics are shown in Table 2.1. Many studies have shown that Gram negative rods belonging to the *Bacteroides fragilis* group are the numerically predominant bacteria in the colon, and it has been estimated that they can account for up to 30% of the total anaerobic count (Macy and Probst 1979). The other main groups consist of an assortment of Gram positive rods and cocci. Chief amongst these are the bifidobacteria, which have been reported to

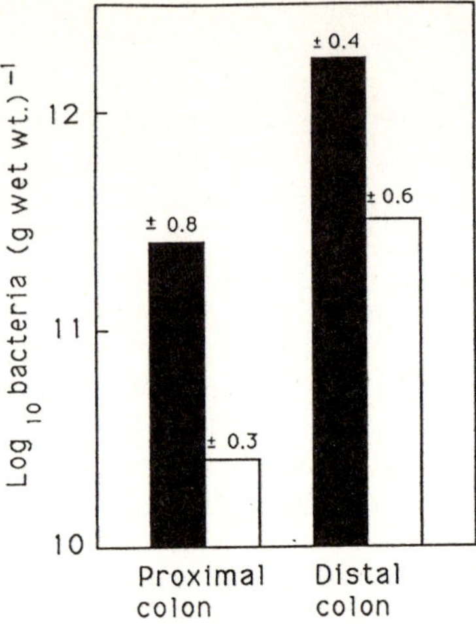


**Fig. 2.2.** Comparison of direct microscope counts (closed bars) and viable counts (open bars) of anaerobic bacteria in different regions of the colon. Results are mean values ± SD, N = 5. (G.T. Macfarlane, unpublished results).

**Table 2.1.** Description of the numerically predominant anaerobes that occur in the large intestine

| Bacteria | Description | Numbers reported in faeces $Log_{10}$ (g dry wt.)$^{-1}$ Mean | Range | Nutrition | Principal metabolic products |
|---|---|---|---|---|---|
| Bacteroides | Gram negative rods | 11.3 | 9.2 - 13.5 | Saccharolytic | A, P, S |
| Eubacteria | Gram positive rods | 10.7 | 5.0 - 13.3 | Saccharolytic, some amino acid fermenting species | A, B, 1 |
| Bifidobacteria | Gram positive rods | 10.2 | 4.9 - 13.4 | Saccharolytic | A, 1, f, e |
| Clostridia | Gram positive rods | 9.8 | 3.3 - 13.1 | Saccharolytic and amino acid fermenting species | A, P, B, L, e |
| Lactobacilli | Gram positive rods | 9.6 | 3.6 - 12.5 | Saccharolytic | L |
| Ruminococci | Gram positive cocci | 10.2 | 4.6 - 12.8 | Saccharolytic | A |
| Peptostreptococci | Gram positive cocci | 10.1 | 3.8 - 12.6 | As for the clostridia | A, L |
| Peptococci | Gram positive cocci | 10.0 | 5.1 - 12.9 | Amino acid fermenters | A, B, L |
| Anaerobic streptococci | Gram positive cocci | 10.3 | 7.0 - 12.3 | Saccharolytic | L, a |
| Methanobrevibacters | Gram positive coccobacilli | 8.8 | 7.0 - 10.5 | Chemolithotrophic | $CH_4$ |
| Desulfovibrios | Gram negative rods | 8.4 | 5.2 - 10.9 | Various[a] | A |

[a] These bacteria can grow chemolithotrophically on $H_2$ and $CO_2$ fermentatively, or oxidatively on some organic acids using $SO_4^{2-}$ as a terminal electron acceptor.
[b]With the exception of the methanogenic bacteria (Miller and Wolin 1982; Jones *et al* 1987) and the sulphate-reducing bacteria (Gibson *et al* 1988a; Gibson 1990), the cell count results are taken from Finegold *et al* (1983), and the fermentation product information from Holdeman *et al* (1977).
[c]A = acetate, P = propionate, B = butyrate, L = lactate, f = formate, e = ethanol

constitute as much as 25% of the gut flora, with *B. adolescentis* and *B. longum* predominating in adults (Mitsuoka 1984; Scardovi 1986).

Somewhat surprisingly, the relative numbers and types of bacteria that have been found by different investigators around the world are generally similar, although this may be an artefact of the cultural methods that are used in their detection, which are strongly influenced by the approach of the clinical microbiologists. For example, many bacterial forms that can be observed microscopically in gut contents cannot be found in subsequent isolation procedures (Croucher *et al.* 1983). Whilst this may be explained by the fact that the cellular morphology of some bacteria is strongly influenced by nutritional availability and environmental stress, it is also likely that many gut species remain to be identified. A similar situation has been noted in studies on the gut microflora of rodents, where despite extensive research over many years, it is considered that only about 10% of the bacteria have been characterised (Wilkins 1981).

The existence of distinct mucosal and lumenal bacterial populations in the human colon has been addressed by many groups, since in many animals including chickens (Lee 1980), ruminants (Wallace *et al.* 1979) and termites (Breznak and Pankratz 1977), a specific microflora has been found growing in association with epithelial surfaces. Although there is some evidence for independent mucosal populations in man, in that certain organisms with distinct morphological characteristics can be visualised in situ on the mucosa, but cannot be seen or cultured from lumenal contents (Lee *et al.* 1971; Croucher *et al.* 1983), some workers have reported that mucosal populations are generally similar to those present in the lumen (Nelson and Mata 1970).

Viewed as a whole, the gut microflora is widely considered to be a remarkably stable entity within an individual (Bornside 1978) but this should not be taken to mean that a static situation exists in which all of the component populations are unchanging. Indeed, it is probable that at the level of individual species, considerable variations in cell population densities occur over long periods. That this is so was indicated in an investigation by Meijer-Severs and Van Santen (1986), who studied ten human volunteers over a 12-month period and found up to 1000-fold differences in total anaerobe counts in some subjects during the course of the investigation.

The composition of the gut flora appears to be influenced to some degree by diet, age and geographical considerations, but these factors are not thought to be particularly significant (Finegold *et al.* 1974; Holdeman *et al.* 1976; Finegold and Sutter 1978), at least as far as the commonly studied bacterial groups are concerned. However, Aries *et al.* (1969) and Hill *et al.* (1971) found higher levels of bifidobacteria and bacteroides in the faeces of British and Americans compared to Japanese, Indians and Ugandans, whereas the latter group had higher counts of enterobacteria and streptococci.

Few investigators have looked at changes in the gut flora associated with ageing, although Gorbach *et al.* (1967) reported that older people have reduced levels of bifidobacteria and increased enterobacterial populations.

## Factors Affecting the Structure and Stability of the Gut Flora

The large intestine is a dynamic environment, yet the climax communities of microorganisms that it contains are by definition relatively stable in their composition. Although it is usually convenient to consider the gut ecosystem as a single entity, due to the heterogeneous nature of gut contents, individual bacteria exist in a multiplicity of different microhabitats and metabolic niches.

As mentioned earlier, a variety of host-associated factors may potentially affect the make-up of the flora, including environmental and dietary influences, transit time of colonic contents and pH, however interactions between the bacteria themselves are probably the ultimate determinants of species composition. As a general rule, such relationships are strongly dependent on the type of environment in which the bacteria exist.

Microbial interactions in the large bowel can occur between organisms in the same or in different habitats and metabolic niches. A variety of different potential environments can be readily identified in gut contents such as those associated with the mucosa or lumen, which were previously discussed. In the lumen, bacteria may be "free-living", present as microcolonies, or associated with particulate materials (Englyst *et al.* 1987).

Many different types of relationship are possible between colonic bacteria, including neutralism, where bacterial populations grow together but have no significant effect on each other; commensalism, where one species is stimulated by a second, which is unaffected by the growth and activities of the first; amensalism or antagonism, in which one population is repressed by inhibitory substances produced by another, and symbiosis, where two species have an obligate dependence on each other.

However, the ability to compete for limiting nutrients and possibly for adhesion sites on food particles or on the colonic mucosa, is likely to be the most important factor that determines the composition of the gut flora, with species that are unable to compete successfully being rapidly eliminated from the system. This leads to a consideration of the substrates that are available for the growth of bacteria in the large intestine.

## Growth Substrates

One of the main reasons for the great variety of bacterial species found in the human colon is thought to be the multiplicity of different carbon sources of both host and dietary origin, to which the bacteria potentially have access (Freter 1983).

The principal substrates are shown in Table 2.2. Starches that for various reasons are resistant to the action of pancreatic amylase, can be degraded by bacterial amylases (Englyst and Macfarlane 1986) and these substances, together with plant cell wall polysaccharides (dietary fibre) are quantitatively of greatest significance. Dietary fibre can be chemically separated into cellulose and non-cellulosic polysaccharides, which includes pectins, hemicelluloses and other molecules such as gums and mucilages. These substances are mainly composed of glucose, galactose, xylose, arabinose and uronic acids (Cummings and Branch 1986; Englyst *et al.* 1987). Typically, about 50% of the amorphous cellulose in vegetables is fermented in the colon and this value rises to about 80% if all plant cell wall polysaccharides are taken into account (Cummings 1983).

Other carbohydrate sources include oligosaccharides and a variety of sugars and non absorbable sugar alcohols (Calloway and Murphy 1968; Tadesse *et al.* 1980).

Host-produced carbohydrates include gastric, small intestinal and colonic mucins. These are glycoproteins secreted by goblet cells in the gastrointestinal tract, where they form a protective gel covering the mucosal surface (Smith and Podolski 1986). Up to 85% of the weight of some mucin molecules may be carbohydrate (Smith and Podolski 1986), in the form of oligosaccharide side chains that are comprised of varying amounts of fucose, galactose, hexosamines and sialic acids (Hoskins and Boulding 1981).

**Table 2.2.** Quantitative estimates of substrate availability for bacteria growing in the large intestine in individuals consuming Western diets

| Substrate | Amount ($gd^{-1}$) | Reference |
|---|---|---|
| *Dietary origin* | | |
| Resistant starches | 8 - 40 | Englyst and Cummings (1987) |
| Non-starch poysaccharides | 8 - 18 | Bingham *et al.* (1990); Englyst *et al.* (1988; 1989) |
| Oligosaccharides | 2 - 8 | Bond *et al.* (1980); Wiggins (1984) |
| Unabsorbed sugars | 2 - 10 | Launiala (1968a,b); Tadesse *et al.* (1980); Hosoya *et al.* (1988) |
| Proteins and peptides | 12 | Gibson *et al.* (1976); Chacko and Cummings (1988) |
| *Endogenous sources* | | |
| Pancreatic enzymes and other gut secretions | 4 - 6 | Kuknal *et al.* (1965) |
| Small bowel mucins | 2 - 3 | Stephen *et al.* (1983) |
| Exfoliated epithelial cells | Unknown | - |
| Colonic mucins | Unknown | - |
| Bacterial secretions and lysis products | Unknown | - |

Other fermentable substrates of endogenous origin include tissue mucopolysaccharides such as hyaluronic acid and chondroitin sulphate, which are respectively composed of glucuronic acid and N-acetyl glucosamine and glucuronic acid and N-acetylgalactosamine (Jeanloz 1970). Epithelial cells may be important substrates for colonic bacteria, since turnover of colonic epithelium is rapid, with columnar absorptive cells and goblet cells having a half-life of about six days (Christensen 1991). Cell death and subsequent destruction by the microflora takes place rapidly after exfoliation (Shamsuddin 1989). Some degree of the significance of host substrates can be inferred from an interesting study by Miller *et al.* (1984) who showed that anaerobic bacteria, including methanogens grew in an isolated segment of sigmoid colon of a patient, in the absence of any supply of dietary substrate.

Many factors influence the degradation of polymeric carbohydrates in the large bowel, including physical consistency of gut contents and particle size (Brodribb and Groves 1978; Heller *et al.* 1980), the solubility of the substrate (Englyst *et al.* 1987) and its occurrence in complexes with other materials (van Soest 1975). The degree of lignification of plant cell material is also important, with highly lignified substances being degraded most slowly (Southgate *et al.* 1976). Intestinal transit time also affects the digestibility of certain polymers including cellulose and hemicelluloses, with longer transits facilitating more extensive breakdown (Hummel *et al.* 1943; Stephen *et al.* 1987).

The role of proteins and their peptide and amino acid hydrolysis products as growth substrates for human gut bacteria has often been overlooked or ignored, but as can be seen from Table 2.2, large quantities are present in the colon, including proteins in dietary residues, tissue proteins such as collagen, serum albumins and proteins produced by the bacteria themselves. Quantitatively however, pancreatic secretions containing a variety of hydrolytic enzymes (proteases, peptidases, amylase, lipase, DNAase, RNAase) are probably amongst the most important sources of organic nitrogen available for the bacteria. Thus, it can be seen that through the degradative activities of its bacteria, the colon is active in reclaiming cell carbon and nitrogen which would otherwise be lost. About half (0.5 to 4.0 $gNd^{-1}$) of the nitrogen losses from the small bowel are absorbed as bacterial metabolites (Gibson *et al.* 1976; Wrong *et al.* 1981), indicating the significance of this recycling process. Of the remaining nitrogen that is excreted in faeces, between 70% - 80% is incorporated in bacterial cell mass (Stephen and Cummings 1979)

# Carbohydrate Metabolism

## Polysaccharide Breakdown

Carbohydrate availability is an important factor limiting bacterial growth in the large intestine. As has been seen, many different types of substrate are present in the gut at any one time, but their relative concentrations will be continuously changing as they are broken down, replenished or replaced. Substrates utilised by the predominant gut anaerobes are shown in Table 2.3. Of note is the great nutritional versatility of the bacteroides.

**Table 2.3.** Polysaccharides variously utilised by gut anaerobes

| Bacteria | Polysaccharide degraded |
|---|---|
| Bacteroides | Pectin, cellulose, xylan, guar gum, mucins, heparin, chondroitin sulphate, starch |
| Bifidobacteria | Pectin, starch, mucin, gum arabic, xylan |
| Ruminococci | Mucins, guar gum |
| Eubacteria | Pectin, starch |

Human intestinal bacteria, particularly the bacteroides, are highly adapted to growth on complex polymerised carbon sources. They are able to utilise these substances through their ability to produce a variety of largely inducible polysaccharidases and glycosidases (Salyers and Leedle 1983; Macfarlane *et al.* 1990) which are not synthesised by the host,

The activites of these hydrolytic enzymes in faeces were studied by Englyst *et al.* (1987), who found that the majority of enzyme activity was associated with the bacteria, with a smaller proportion occurring extracellularly (Table 2.4). Interestingly, the activities of some of the enzymes involved in breaking down some of the less soluble polymers such as xylanase, arabinogalactanase, ß-xylosidase and α-arabinofuranosidase, were markedly higher in bacteria that were growing attached to particulate material, compared to those which were "free living." This data indicates that either different bacterial populations were growing attached to particles in the gut, or alternatively, that particle-associated bacteria were expressing increased enzyme levels due to the presence of comparatively higher amounts of inducer substances at the particle surface. These results suggest that adherent bacteria may be particularly important in digesting insoluble plant cell wall materials in the gut.

The complete destruction of a complex polysaccharide in the large bowel is likely to be dependent on the activities of a number of different hydrolytic enzymes, that can breakdown the backbone and side chains of the polymer. Although some colonic bacteria can produce a number of different polysaccharidases and glycosidases, which in theory would enable them to extensively digest a heterogeneous polysaccharide (Pettipher and Latham 1979; Berg 1981; Macfarlane *et al.* 1990), it is likely that polymer degradation is a cooperative activity with enzymes from many different bacteria taking part in the process.

**Table 2.4** . Polysaccharidase and glycosidase activities in different fractions of human faeces obtained from five individuals[a]

| Substrates | Fraction | | |
|---|---|---|---|
| | Extracellular | Extract from washed bacterial cells | Extract from washed bacteria removed from particulate materials by surfactant treatment |
| Polysaccharidases[b] | | | |
| Starch | 48.20 ± 11.34 | 4.25 ± 1.16 | 3.96 ± 1.62 |
| Xylan | ND | 0.82 ± 0.18 | 1.40 ± 0.47 |
| Pectin | 0.66 ± 0.15 | 1.09 ± 0.29 | 1.14 ± 0.32 |
| Arabinogalactan | 0.19 ± 0.05 | 0.43 ± 0.13 | 0.79 ± 0.16 |
| Galactomannan | ND | 0.06 ± 0.02 | ND |
| Glycosidases[c] | | | |
| p-Nitrophenyl α-L-arabinofuranoside | 65 ± 34 | 1598 ± 522 | 2782 ± 573 |
| p-Nitrophenyl ß-D-xylopyranoside | 102 ± 69 | 955 ± 153 | 1986 ± 308 |
| p-Nitrophenyl α-D-galactopyranoside | ND | 930 ± 125 | 832 ± 66 |
| p-Nitrophenyl ß-D-galactopyranoside | 434 ± 168 | 2570 ± 520 | 3231 ± 491 |
| p-Nitrophenyl ß-D-galacturonide | ND | 4 ± 3 | 86 ± 47 |
| p-Nitrophenyl ß-D-glucuronide | ND | 130 ± 38 | 44 ± 18 |
| p-Nitrophenyl ß-D-glucopyranoside | 50 ± 35 | 1581 ± 474 | 1293 ± 261 |
| p-Nitrophenyl α-D-glucopyranoside | 113 ± 73 | 511 ± 76 | 586 ± 149 |
| p-Nitrophenyl α-D-mannopyranoside | 78 ± 44 | 16 ± 9 | 28 ± 14 |
| p-Nitrophenyl ß-D-mannopyranoside | 20 ± 8 | 52 ± 23 | 22 ± 13 |

Data are from Englyst *et al.* (1987)
[a] Results are ± SEM
[b] mmol reducing sugar released $h^{-1}$ mg $protein^{-1}$
[c] nmol p-nitrophenol released $h^{-1}$ mg $protein^{-1}$
ND, not detected

In contrast to the depolymerisation of complex polysaccharides, there is undoubtedly considerable competition between colonic bacteria for the hydrolysis products of polymer breakdown, since there are substantial populations of saccharolytic bacteria in the gut that are unable to digest polysaccharides by themselves. These must grow by cross-feeding on carbohydrate fragments produced by the polysaccharide degraders.

Studies with pure cultures of intestinal bacteria indicate how this can occur. Macfarlane and Gibson (1991b) grew *Bacteroides ovatus*, one of the predominant polymer degrading species in the gut, in continuous culture, on a mixture of starch and arabinogalactan. Enzyme measurements showed that although polysaccharidase and glycosidase activities were primarily cell-associated, substantial amounts of starch and arabinogalactan-derived oligosaccharides and sugar monomers accumulated in the fermenters, during both C and N-limited growth. Thus, utilisation of polysaccharides by this bacterium appears to be a relatively inefficient process, and if similar events occur in the large intestine, the partially digested carbohydrate would be rapidly utilised by other species. These experiments also demonstrated the metabolic flexibility of the organism in that it could simultaneously ferment a number of different carbohydrate sources. In earlier studies, mixed populations of faecal bacteria were also found to co-utilise polymerised substrates when they were present in a mixture (Englyst *et al.* 1987).

A number of factors are likely to affect the utilisation of fermentable polysaccharides in the colon. Important among these is solubility: the more soluble substrates, being more accessible to hydrolytic enzymes, are likely to be degraded more rapidly. This was demonstrated by Englyst *et al.* (1987) and Gibson *et al.* (1990b) in studies which showed that different polysaccharides are broken down by gut bacteria at different rates. Their work showed (Fig.2.3) that the relatively insoluble substrate xylan was degraded more slowly than more soluble substrates such as starch and pectin. As will be seen later, this may influence the types of fermentation products that are formed.

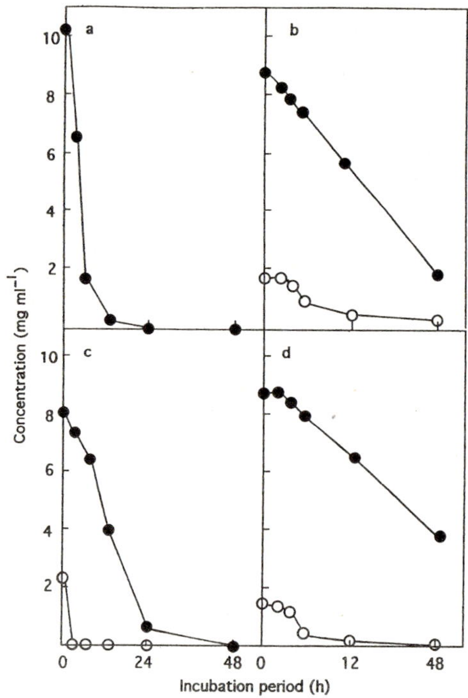


**Fig. 2.3.** Breakdown of starch (**a**), arabinogalactan (**b**), pectin (**c**) and xylan (**d**) by mixed populations of intestinal bacteria. Closed symbols are backbone sugars, open symbols are side chain sugars. Results are from Englyst *et al.* (1987).

## Carbohydrate Fermentation

Fermentation of carbohydrate by anaerobic bacteria growing in the large intestine consists of a series of energy-yielding reactions that do not involve respiratory chains in which molecular oxygen is the terminal electron acceptor. Instead, the electron acceptor in these processes is usually one of the products of fermentation. In some fermentations, ATP is formed by substrate level phophorylation reactions, whereas in others respiratory chains may be involved. In studies on the efficiency of fermentation in the human colon, cell yields were estimated to be in the region of 25 g - 36 g bacteria per 100 g of carbohydrate utilised (Cummings 1987), which is comparable to values obtained with rumen bacteria (Hespell and Bryant 1979).

With the notable exception of the bifidobacteria, the majority of saccharolytic anaerobes in the colon use the Embden-Meyerhoff-Parnas pathway to ferment carbohydrates (Wolin and Miller 1983). A summary of the principal carbohydrate fermentations is shown in Fig. 2.4. The main end-products of metabolism are the short chain fatty acids (SCFA) acetate, propionate and butyrate (Cummings 1981; Cummings *et al.* 1987), but a variety of other metabolites are also produced including electron sink products such as lactate, ethanol, hydrogen and succinate (Fig. 2.5). These electron sink products are formed to maintain redox balance during fermentation, enabling oxidation and recycling of reduced pyridine nucleotides. In the context of the ecosystem as a whole, they act as fermentation intermediates, since they are further metabolised to SCFA by other species. The production of these substances in fermentation therefore contributes to the diversity of bacteria found in the colon.

Fermentation reactions are controlled by the need to maintain redox balance, through the reduction and reoxidation of pyridine nucleotides. This in turn influences the flow of carbon through the bacteria, the energy yield obtained from a particular substrate and the fermentation products that are formed.

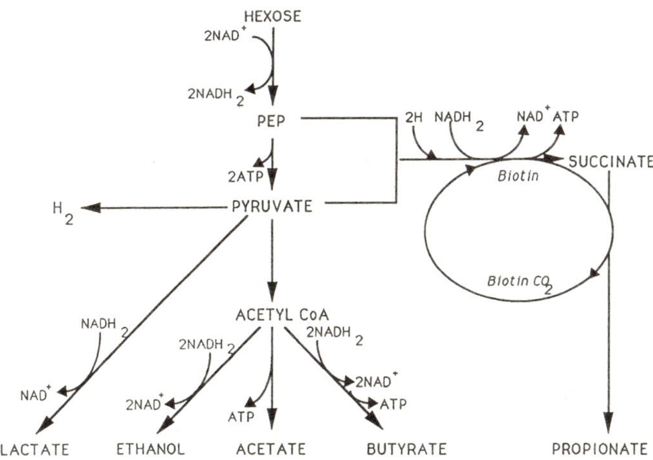


**Fig. 2.4.** Generalised scheme of carbohydrate fermentation by intestinal bacteria.

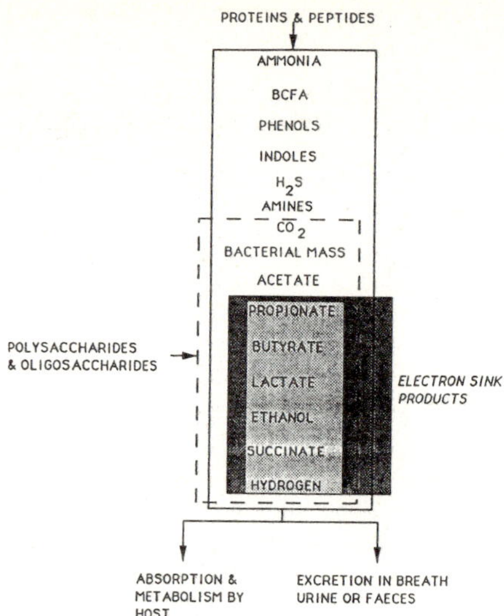


**Fig. 2.5.** Relative contributions of carbohydrate and protein to different fermentation products in the colon.

The effect of carbon availability on fermentation product formation can be seen in Fig. 2.6, where measurements were made of bacterial metabolites in gut contents obtained from different regions of the colon. Marked differences are evident in a number of the parameters studied in the carbohydrate-rich proximal colon and the carbohydrate-depleted distal colon. For example, total SCFA decrease from 142 mmol $kg^{-1}$ in the proximal colon to 96 mmol $kg^{-1}$ in the distal colon. This is accompanied by an increase in pH. Although the amounts of SCFA vary in different regions of the gut, the molar ratios of acetate, propionate and butyrate show little change (proximal colon 57:22:21; distal colon 57:21:22) (Cummings *et al.* 1987). Concentrations of lactate, ethanol and succinate are also lower in the distal gut, reflecting reduced substrate levels. The increased capacity for SCFA production in the proximal colon was further demonstrated by Macfarlane *et al.* (1992a), who incubated gut contents *in vitro*, in the absence of exogenous carbon sources and found that SCFA formation was three times higher in the proximal colon than in the distal gut. These measurements provide a good example of how the physiological and anatomical structure of different parts of the colon influences the availability of substrate for bacteria in the gut and hence, their activities.

**Table 2.5.** Factors affecting SCFA production from polysaccharides by colonic bacteria

| | |
|---|---|
| 1 | Chemical composition of the fermentable substrate |
| 2 | Amount of substrate available |
| 3 | Rate of depolymerisation of substrate |
| 4 | Substrate specificities and preferences of individual gut species |
| 5 | Fermentation strategies of substrate utilising species |
| 6 | Presence of inorganic terminal electron acceptors |

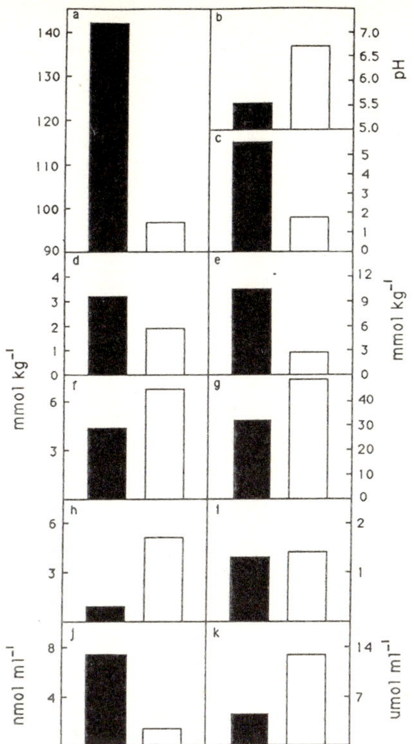


**Fig. 2.6.** Regional differences in concentrations of bacterial metabolites in the large intestine. Closed bars correspond to the proximal colon and open bars to the distal colon. **a**, total SCFA; **b**, pH; **c**, lactate; **d**, succinate; **e**, ethanol; **f**, branched chain fatty acids; **g**, ammonia; **h**, total phenols; **i**, sulphide; **j**, hydrogen; **k**, methane. Results are from Macfarlane *et al.* (1992a).

A variety of factors can potentially affect the outcome of fermentation reactions in the colon, some of which are summarised in Table 2.5. The chemical composition of the substrate markedly influences the fermentation products that are formed, as evidenced by the work of Englyst *et al.* (1987) who showed that acetate and butyrate were the principal SCFA produced from starch by faecal bacteria, whereas acetate was the main fermentation product from pectin and xylan. Other studies have shown lactate is an important intermediate of starch fermentation (see Fig. 2.7) but not fermentation of plant cell wall polysaccharides (Macfarlane and Englyst 1986; Etterlin *et al.* 1992). The relationship between starch and lactate is further demonstrated in Fig. 2.8 where polymerised glucose (starch) and butyrate were measured in caecal contents taken from nine persons who had died suddenly. The results show a strong correlation between starch and lactate in these samples although lactate concentrations, together with those of ethanol and succinate, are generally kept relatively low in the bowel (see Fig. 2.3), due to cross-feeding.

The quantity of substrate that is available for intestinal bacteria influences the way that it will be utilised. Obviously, if large amounts of substrate are present, more bacteria will have access to it, which will increase the type of fermentation product that can be formed. However, it will also affect the outcome of fermentation by individual

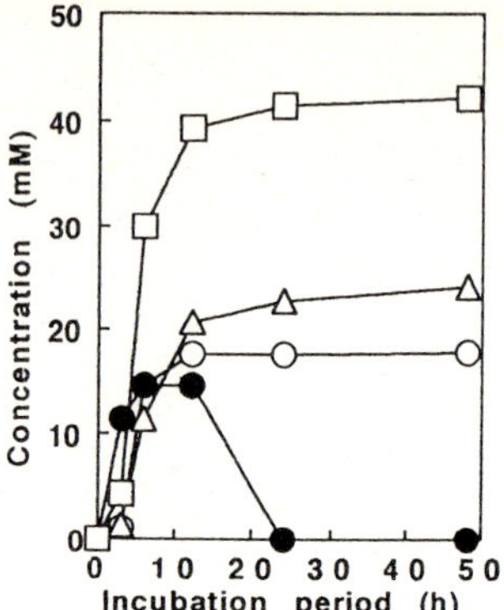


**Fig. 2.7.** Lactate and SCFA production by mixed populations of intestinal bacteria grown on starch in batch culture. Lactate (●), acetate (□), propionate (O) and butyrate (Δ). Data are taken from Macfarlane and Englyst (1986).

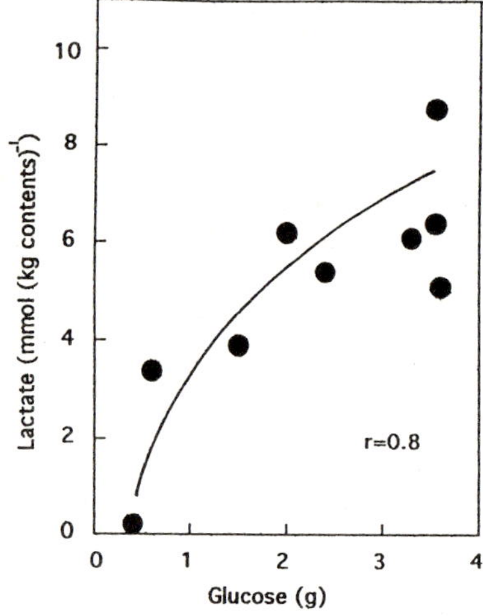


**Fig. 2.8.** Relationship of polymerised glucose (starch) and lactate concentrations in human caecal contents (N = 9). J.H. Cummings (unpublished data).

species. This is clear from physiological studies with some clostridia and certain members belonging to the *Bacteroides fragilis* group. In lactate producing clostridia such as *C. perfringens*, high substrate concentrations result in lower acetate production and increased lactate formation (Allison and Macfarlane 1989b), whereas in saccharolytic butyrate producing clostridia, increasing the carbohydrate supply results in more butyrate and less acetate being formed, because butyrate, like lactate with *C. perfringens*, is used as a sink to dispose of excess reducing power (Rogers 1986).

In *Bacteroides ovatus*, which produces acetate, propionate and succinate as fermentation products, continuous culture studies have demonstrated that acetate formation is maximal during C-limited growth, whereas during growth under conditions of carbon-excess, more succinate is synthesised together with some lactate (Macfarlane and Gibson 1991b).

Increasing carbohydrate availability also influences fermentation in other bacteria and has been observed to increase alcohol formation in a number of Gram positive anaerobes (Turton *et al.* 1983). Substrate availability will also affect bacterial growth rates which can in turn influence the synthesis of depolymerising enzymes (Macfarlane and Gibson 1991a).

Because the majority of carbohydrate entering the large bowel is in the form of polysaccharides, it is likely that the rate at which these substances can be depolymerised controls the rate at which fermentable substrate becomes available for assimilation by the bacteria, and for reasons given above, this will affect fermentation product formation.

In the colon, bacteria occupy a variety of metabolic niches. As a result, the relative substrate specificities, preferences and fermentation strategies of individual species will play a role in determining the types of fermentation product that can be formed from a given substrate. This can be seen in Tables 2.1 and 2.3. Taking bacteroides as an example, during growth on any one of a number of polysaccharides, the principal fermentation products will be acetate, propionate and succinate. As we have seen however, the relative amount of each of these products that can be formed during fermentation is subject to other controls. Another factor which will influence fermentation product formation, is the degree to which cross-feeding species have access to the substrate.

Substrate preferences may affect the way in which different carbohydrates are utilised. They have been identified in some rumen anaerobes (Russell and Baldwin 1978), together with human colonic bifidobacteria (Degnan and Macfarlane 1991), and may be of some significance since they will have the effect of compartmentalising competition between bacteria for substrates. This would result in the formation of different nutritional niches, with distinct component populations, each competing for their own preferred range of substrates.

The final factor affecting fermentation product formation by gut bacteria to be considered here, is the availability of inorganic electron acceptors. Studies have indicated that the presence of $NO_3^-$ can result in more oxidised fermentation products being formed. Allison and Macfarlane (1988) found that diversion of electrons to $NO_3^-/NO_2^-$ reduction resulted in mixed populations of gut bacteria producing more acetate and less butyrate. Similar effects have been shown to occur in some saccharolytic clostridia (Keith *et al.* 1982), whilst studies with pure cultures of intestinal propionibacteria demonstrated that during lactate fermentation, more acetate and less propionate was formed when $NO_3^-$ was present (Allison and Macfarlane 1989c).

# Metabolism of Proteins and Peptides

## Proteolysis

Quantitatively, most of the nitrogen entering the human colon does so in the form of proteins and peptides. Concentrations of ammonia, urea and free amino acids are low in ileal effluent (Chacko and Cummings 1988) and bacteria must therefore depolymerise these macromolecules to make available the carbon and nitrogen that they contain.

The initial step in this process involves hydrolysis of the proteins to smaller peptides, by proteolytic enzymes. Some gut bacteria prefer to assimilate amino nitrogen in the form of peptides ( Pittman *et al.* 1967; Wright 1967), whereas others utilise free amino acids (Wahren and Holme 1973; Barker 1981). Proteolysis is therefore a crucial process in the large gut in that it provides nitrogen for growth by saccharolytic bacteria, and amino acids for fermentation by asaccharolytic species.

The human large intestine is a highly proteolytic environment containing a complex mixture of bacterial and pancreatic proteases (Macfarlane *et al.* 1986). Bacterial enzymes contribute considerably to overall protease activity in the colon, especially in the distal gut, and play an important role in digesting pancreatic enzymes which are the preferred organic N-sources for some gut bacteria (Macfarlane and Macfarlane 1991). One manifestation of this is that total protease activity decreases as gut contents move distally through the bowel (Macfarlane *et al.* 1989). The extent of the contribution of bacterial enzymes to faecal protease activity was shown by Macfarlane *et al.* (1988b), who compared faecal samples from 10 healthy volunteers, with samples from an individual whose pancreas had been surgically removed. The results showed that although faecal proteolysis was generally lower in this individual, it still remained within the normal range of activities found in healthy persons.

This study further demonstrated that faecal proteolysis, which was found extracellularly and associated with the bacteria and particulate materials, was of a substantially different character from pancreatic protease activities in small intestinal contents. Inhibitor experiments showed that faecal proteases primarily consisted of a mixture of serine, cysteine and metalloproteases, whereas pancreatic proteases (trypsin, chymotrypsin, elastase) are of the serine type.

**Table 2.6.** Hydrolysis of different substrates by proteases in small intestinal effluent and faeces[a]

| | Rate of hydrolysis[b] | | |
|---|---|---|---|
| | Casein | BSA | Collagen |
| Ileal effluent | 1214 ±546 | 0 | 281 ± 35 |
| Faeces | 20 ± 2 | 2 ± 1 | 6 ±2 |

[a]Results are from Gibson *et al* (1989) and are mean values from five persons ±SD
[b]mg protein $h^{-1}$ (gram wet wt contents)$^{-1}$

Faecal proteolysis differs both qualitatively and quantitatively from that in the small bowel. Table 2.6 shows the results of an investigation by Gibson *et al.* (1989) who compared the ability of pancreatic proteases in ileal effluent and the proteolytic activity in faeces to digest different protein substrates. It can be seen that there is substantial reduction in protease activity in faeces compared to small gut contents, and importantly, that different proteins are digested in different ways. For example, casein was broken down more rapidly than collagen by both faecal and small intestinal proteases, whereas BSA was not hydrolysed by pancreatic enzymes at all, but was digested by faecal proteases. This would suggest that pancreatic proteases may not be particularly efficient at breaking down some highly globular proteins in the small bowel.

A wide variety of bacteria that occur in the large intestine have been reported to be proteolytic, including saccharolytic (Allison and Macfarlane 1989a) and asaccharolytic clostridia (Allison and Macfarlane 1990; Macfarlane and Macfarlane 1992), *Prevotella* spp. (Hausmann and Kaufman 1969), lactobacilli (El-Soda *et al.* 1978), fusobacteria (Wahren *et al.* 1971), streptococci (Macfarlane *et al.* 1986), propionibacteria (Ingram *et al.* 1983) and most importantly species belonging to the *Bacteroides fragilis* group (Gibson and Macfarlane 1988; Macfarlane and Macfarlane 1991).

The proteases formed by intestinal bacteria may be true extracellular enzymes, as is the case with clostridia and propionibacteria, or cell-associated, as occurs in lactobacilli. Proteases produced by members of the *B. fragilis* group are interesting in that they are cell-bound during exponential growth in batch culture, with large amounts accumulating intracellularly, but as growth slows, these intracellular enzymes are released into the culture medium (Macfarlane *et al.* 1992b).

## Amino Acid Fermentation

Substantial populations of amino acid fermenting bacteria are present in the large intestine. Many species, including the peptococci, acidaminococcus, veillonella, together with some fusobacteria, eubacteria and clostridia either do not, or only weakly ferment carbohydrates (Holdeman *et al.* 1977). These organisms and similar bacteria therefore form a metabolically and ecologically distinct group in the large bowel.

The fermentation of amino acids by intestinal bacteria consists of a range of oxidation and reduction reactions. Many different types of electron acceptors can be used, including keto acids, other amino acids, unsaturated fatty acids and $H_2$ (Barker 1981). The fermentation of some amino acids such as alanine, leucine, glutamate and valine generates $H_2$ (Barker 1981; Nagase and Matsuo 1982), whilst in Stickland reactions, $H_2$ is generated from one amino acid by the reduction of another (Nisman 1954). Because some amino acid fermentations carried out by anaerobic bacteria only occur at low partial pressures of $H_2$ (Stams and Hansen 1984), the occurrence of interspecies $H_2$-transfer reactions with methanogenic, sulphate-reducing or acetogenic bacteria may be of considerable significance.

Although many of the end products of the catabolism of carbohydrates and amino acids are similar, compared to the fermentation of carbohydrates, a wider range of metabolic products are formed from the metabolism of amino acids. These include ammonia, amines, phenols, indoles, organic acids, alcohols and the gases $H_2$ and $CO_2$ (see Fig. 2.5). Quantitatively however, SCFA are the predominant organic end products of dissimilatory amino acid metabolism (Macfarlane and Allison 1986). These authors showed that during the *in vitro* digestion of protein by faecal bacteria, about 30% of the substrate was converted to SCFA, principally acetate, propionate, butyrate and the branched chain fatty acids (BCFA) isobutyrate, 2-methylbutyrate and isovalerate. These BCFA are mainly formed by the metabolism of branched-chain amino acids (valine, isoleucine and leucine respectively) and considerable amounts can be produced in Stickland reactions where the branched chain fatty acids typically serve as electron donors and give rise to SCFA that are one carbon shorter than the parent amino acid (Elsden and Hilton 1978; Andreesen *et al.* 1989).

In a recent study, Macfarlane *et al.* (1992c) used the occurrence of BCFA in different regions of the colon to estimate the contribution of proteins and peptides to SCFA production. They found that BCFA comprised 3.4% and 7.5% of total SCFA in the proximal and distal colons respectively, which suggested that the fermentation of protein could account for about 17% of SCFA produced in the caecum and 38% of SCFA in the sigmoid/rectum. The results also showed that protein breakdown became quantitatively more significant in the distal colon.

Unlike carbohydrate metabolism, many of the products of protein breakdown, or putrefaction, are toxic to the host. Ammonia concentrations vary markedly in faeces, ranging from 3 mM to 44 mM (Wilson *et al.* 1968). In the large bowel, concentrations of this metabolite increase with distance from the proximal colon, but the effect is not as marked as with other products of amino acid metabolism, such as phenols and BCFA (see Fig. 2.6).

Although there is still debate as to whether ammonia in the colon results from urea hydrolysis, experiments in which human volunteers were infused with $^{15}$N-labelled urea, have indicated that the majority of ammonia is produced by the deamination of amino acids (Wrong *et al.* 1985).

Ammonia formation from colonic bacteria is of physiological significance to the host, since concentrations as low as 10 mM can alter the morphology and intermediary metabolism of intestinal cells, increase DNA synthesis and affect their lifespan (Visek 1972; Visek 1978). Normally, ammonia produced in the large intestine is transported to the liver, where it is converted to urea, which is excreted in urine. However, in patients with liver disease, ammonia accumulates in body fluids, contributing to the onset of portal-systemic encephalopathy or hepatic coma (Weber *et al.* 1987). Ammonia may also be involved in the initiation of cancer.

Phenolic and indolic compounds are produced by the aromatic amino acids tyrosine, phenylalanine and tryptophan by a variety of deamination, transamination, decarboxylation and dehydrogenation reactions. They are produced by anaerobic bacteria in the gut of man and a number of other animals (Bakke 1969; Spoelstra 1977; Yokohama and Carlson 1979; Ward *et al.* 1987). These processes are carried out by many different types of gut bacteria including clostridia (Elsden *et al.* 1976), bacteroides (Demoss and Moser 1969; Chung *et al.* 1975), enterobacteria (Botsford and Demoss 1972) bifidobacteria (Aragozzini *et al.* 1979) and lactobacilli (Yokohami and Carlson 1981).

Concentrations of phenolic compounds increase markedly in the distal colon as does the relative proportions of p-cresol (4-methylphenol) and phenol to phenyl-substituted fatty acids (Macfarlane *et al.* 1992a), which provides further evidence for the increasing significance of amino acid fermentation in this region of the gut. Phenols and indoles are usually detoxified by either sulphate or glucuronide conjugation by the colonic mucosa (Ramakrishna *et al.* 1989) and daily excretion in urine is in the range 50 - 160 mg (Schmidt 1949). p-Cresol makes up about 90% of the volatile phenols excreted in urine, with phenol and, to a lesser degree, 4-ethylphenol making up the remainder (Schmidt 1949). They do not occur in the urine of germ-free animals.

Although little is known of the physiological and environmental factors that affect the metabolism of aromatic amino acids in anaerobic bacteria, *in vivo* studies by Cummings *et al.* (1979) and *in vitro* studies by Macfarlane *et al.* (1989) have shown that phenol production by faecal bacteria increases with increasing transit time. Cummings *et al.* (1979) also showed that the excretion of phenols is related to both carbohydrate and protein intake, in that increasing the availability of fermentable carbohydrate reduces the production of these metabolites by gut bacteria because of an increase in the requirement by the bacteria for tyrosine for use in biosynthetic reactions.

Phenols and indoles have variously been associated with the initiation of cancer, but they have also been shown to have other effects. For example, skatole which is a metabolite of tryptophan, has been linked to a variety of disease states in man, including malabsorption, anaemia (Horning and Dalgliesh 1958) and schizophrenia (Dalgliesh *et al.* 1958). In the rumen, this metabolite is produced by decarboxylation of indoleacetic acid and is responsible for tryptophan-induced acute bovine pulmonary emphysemsa (Carlson *et al.* 1972). Also in animals, p-cresol production has been implicated as a growth depressant in weanling pigs (Yokohama *et al.* 1982). Studies on the formation of phenolic and indolic compounds by pure cultures of intestinal bacteria are comparatively rare, however clostridia and lactobacilli have been shown to produce both p-cresol and skatole in pure culture (Fellers and Clough 1925; Elsden *et al.* 1976; Yokohama and Carlson 1981).

Amines are major products of bacterial amino acid metabolism. They are principally formed by decarboxylation of amino acids (Gale 1946) and to a lesser extent by N-dealkylation reactions (Johnson 1977), degradation of polyamines (White Tabor and Tabor 1985) or transamination of aldehydes (Drasar and Hill 1974). Histamine, tyramine, piperdine, pyrrolidine, cadaverine, putrescine, 5-hydroxytryptamine and agmatine have all been identified *in vivo* (Drasar and Hill 1974).

As with the formation of phenols and indoles, many different intestinal bacteria are known to produce amines, including bacteroides (Allison and Macfarlane 1989b), clostridia (Andreesen *et al.* 1989), lactobacilli (Sumner *et al.* 1985), enterobacteria (Morris and Boeker 1983) and streptococci (Babu *et al.* 1986).

Although some bacteria may decarboxylate amino acids to maintain intracellular $pCO_2$ (Boeker and Snell 1972), the traditional view is that the formation of these basic molecules constitutes a mechanism whereby bacteria can modify the pH of their environment during growth under acidic conditions (Morris and Fillingame 1974). However, low pH is not always a prerequisite for amine production, since studies with *Clostridium perfringens* have shown that this bacterium optimally forms putrescine, butylamine and propylamine at neutral pH (Allison and Macfarlane 1989b).

Like other small molecular species, amines are probably rapidly absorbed from the gut lumen and oxidised by monoamine and diamine oxidases in the colonic mucosa. However, it is clear that under certain circumstances, amines do enter the general circulation, since they have been reported to be involved in hepatic coma (Phear and Rubner 1956).

Amine excretion in urine has been directly related to protein intake (Irvine *et al.* 1959; DeQuatto and Sjoerdsma 1967), migraine (Anon 1968) and the onset of hypertensive symptoms (Boulton *et al.* 1970). The potential toxicity of amines produced by intestinal bacteria was recognised at the beginning of this century by Metchnikoff (1905) who considered that amines formed by the breakdown of protein in the bowel resulted in enterotoxaemia. Recent studies have shown increased faecal amine levels in infants with gastroenteritis (Murray *et al.* 1986), although it is unclear whether this was a cause or a result of the disease.

A number of amines produced in the colon such as histamine, tyramine, cadaverine and putrescine are pharmacologically active and they can variously function as pressor or depressor substances, acting as stimulators of gastric secretion, or vasodilators (Drasar and Hill 1974).

Amine production by bacteria in the colon is therefore potentially of considerable significance in that it can affect a wide variety of body functions. This is further shown by the fact that amines ingested in foodstuffs have resulted in brain haemorrhage and heart failure (Smith 1980). The association of amines through nitrosamine formation with cancer is discussed in another chapter.

## Physiological Significance of SCFA Production

SCFA are the major anions in the large intestine and are rapidly absorbed through the colonic epithelium (Cummings 1981). Acetate, propionate and butyrate predominate contributing about 50%, 20% and 20% of the total fatty acid pool respectively (Cummings *et al.* 1987). Only about 5% of SCFA produced are excreted in faeces, with the remainder being metabolised by the host (Macfarlane and Cummings 1991). Consequently, measurements of faecal SCFA tell us little about events occurring in the colon.

SCFA production in the large intestine is in the region of 400 mmol per day, with a probable range of 150 - 600 mmol per day (Macfarlane and Cummings 1991). In persons consuming Western-style diets, fatty acids produced in the colon do not make a significant contribution to the hosts energy requirements, since only about 5% (range 3% - 9%) can be met from this source (Cummings *et al.* 1989). They are important in other ways however. Acetate is metabolised in the brain, heart and peripheral tissues (Lindeneg *et al.* 1964; Lundqvist *et al.* 1973; Juhlen-Dannfelt 1977), whereas propionate appears to be mainly cleared by the liver (Cummings *et al.* 1987), although its role as a gluconeogenic precursor is unclear. This fatty acid has been reported to lower blood cholesterol levels in rats (Illman *et al.* 1988) and pigs (Thacker and Bowland 1981), but it does not appear to do so in man (Venter *et al.* 1990).

Butyrate formation by intestinal bacteria is of particular interest, because this SCFA is an essential fuel for the colonic epithelium (Roediger 1980), particularly in the distal gut. Production of butyrate by bacteria in the large bowel may also be important from the viewpoint of preventing colon cancer, since it inhibits DNA synthesis and induces differentiation in a number of human cancer lines, whilst decreasing the proliferation of neoplastic cells and the effects of contact-independent growth (Young 1991). There is consequently great interest in the metabolic functions of butyrate at the present time and it is the subject of considerable research activity.

From a microbiological viewpoint, butyrate formation by gut bacteria is also of interest in that although this fatty acid comprises about 20% of total SCFA in the large bowel, estimates of butyrate producing populations in faeces indicate that these bacteria consitute a relatively small proportion (<1%) of the total anaerobe count (Fig. 2.9). This suggests that they are very active metabolically and the results highlight the observations of Brock (1966) who noted that numbers alone are not necessarily the most important criterion that should be considered when evaluating the metabolic significance of an organism or group of organisms in an ecosystem.

## Hydrogen Metabolism

As discussed earlier, a variety of electron sink products are formed by intestinal bacteria during fermentation reactions in the large intestine, including ethanol, lactate, succinate and $H_2$ (see Fig. 2.5). Although the predominant polysaccharide degrading species such as the *B. fragilis* group and bifidobacteria do not evolve $H_2$ during fermentation, preferring instead to use succinate and lactate as electron sinks, $H_2$ is a major product of carbohydrate and amino acid metabolism in the colon. Measurements of total $H_2$ output (breath and flatus) in man indicate that between 10 and 20 ml of the gas are produced per gram of carbohydrate fermented (Christl *et al.* 1990), but as will be seen, these estimations do not take into account its metabolism by other bacteria.

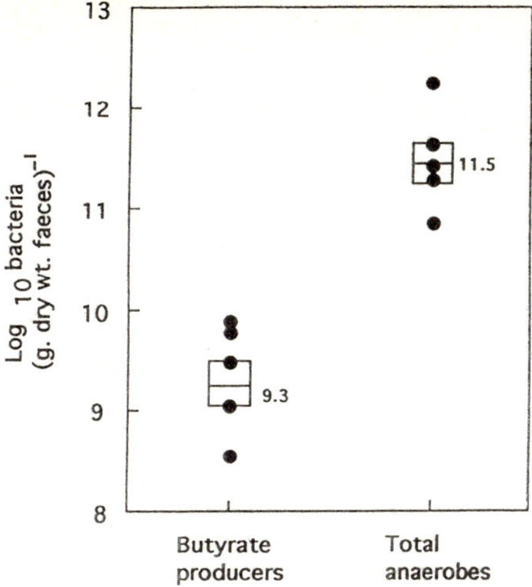


**Fig. 2.9.** Most probable number estimates of butyrate producing bacteria and total anaerobes in human faeces (N = 5). The plot shows mean values and SD. G.T. Macfarlane (unpublished results).

Hydrogen gas is produced by hydrogenase according to the reaction:

$$2H^+ + 2e^- \rightleftharpoons H_2$$

Hydrogenases are synthesised by many different types of fermentative bacteria that occur in the gut. Depending partly on the species involved and partly on $pH_2$, $H_2$ can potentially be formed by a number of routes during carbohydrate breakdown (see Fig. 2.10). Hydrogen generation by saccharolytic bacteria results mainly from the oxidation of pyruvate, formate or reduced pyridine nucleotides (NAD(P)H) and ferredoxins. Pyruvate probably has a central role in $H_2$ metabolism in the large bowel, since it is energetically more favourable to produce $H_2$ from this source than from other sources (Wolin 1976).

Production of $H_2$ from NADH is only thermodynamically possible at low $pH_2$ (Wolin 1974). In ecosystems such as the rumen, $pH_2$ is kept low through interspecies $H_2$ transfer reactions between $H_2$ producers and $H_2$ utilising methanogenic bacteria (Wolin 1979). However, the accumulation of $H_2$ in the human colon is not entirely prevented by either its excretion in breath or flatus, or by the actvities of $H_2$ consuming syntrophs, since it accounts for about 16% of flatus gas (Levitt 1971). It is possible that this may be due to saturation of $H_2$ utilising reactions, however, an alternative explanation may reside in the phenomenon of mass transfer resistance.

Hydrogen is a poorly soluble gas and its transfer across gas-liquid interfaces may be a rate-limiting step in its utilisation by syntrophic species. This has been shown to be an important factor in studies on $H_2$-transfer reactions in thick anaerobic digester sludge (Robinson and Tiedje 1982), which has similar physical characteristics to human gut contents.

The relative spatial locations of $H_2$ producing and consuming bacteria may also influence the fate of $H_2$ in the colonic ecosystem. Evidence for this again comes from studies with sewage sludge from an anaerobic digester, where Conrad *et al.* (1985)

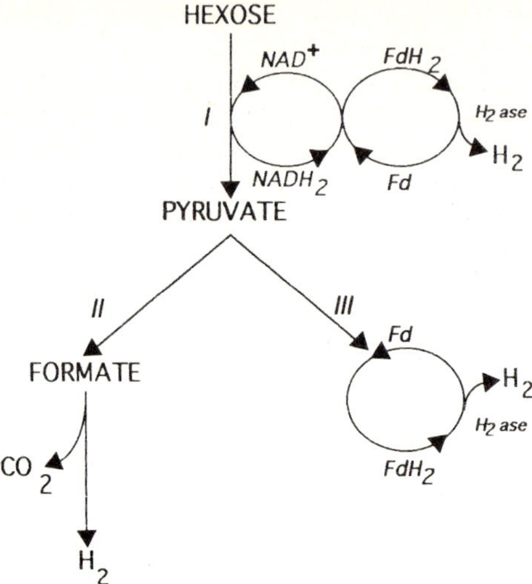


**Fig. 2.10.** Different mechanisms of hydrogen production by human intestinal bacteria. I. $H_2$ formation from $NADH_2$ via NADH: ferredoxin oxidoreductase and hydrogenase. e.g. *Ruminococcus albus*. II.$H_2$ production by pyruvate: ferredoxin oxidoreductase and hydrogenase e.g. butyrate forming clostridia. III. $H_2$ production via formate hydrogen lyase e.g. enterobacteria.

found that over 90% of interspecies $H_2$-transfer reactions occurred between closely associated bacterial populations in aggregates or other microniches. Thus, only a proportion of $H_2$ utilised in this system came from what these authors termed "the common hydrogen pool." If this also occurs in the human colon, it would be of great significance to gastroenterologists and physiologists who use breath $H_2$ measurements to estimate colonic fermentation rates, since any quantitative measurements of $H_2$ excretion would in fact have little relevance to the actual amounts of $H_2$ being produced in the ecosystem.

Hydrogen produced by carbohydrate and amino acid fermenting bacteria in the large bowel can potentially be used as an electron donor and energy source in a variety of reactions including dissimilatory $NO_3^-$ reduction, methanogenesis, dissimilatory sulphate reduction and possibly, acetogenesis. These processes will now be considered in turn.

## Hydrogen as an Electron Donor for Dissimilatory $NO_3^-$ Reduction

Somewhere in the region of 120mg of $NO_3^-$ are consumed by man on a daily basis (Walker 1975; Knight *et al.* 1987). Although the amounts of dietary $NO_3^-$ reaching the colon are a matter of some debate (Bartholomew and Hill 1984; Radcliffe *et al.* 1985), there is evidence that it is secreted into the gastrointestinal tract at several locations (Witter *et al.* 1979; Lee *et al.* 1986) and it seems that macrophages are involved in $NO_3^-$ formation in the colon itself (Beeken *et al.* 1987; Iyengar *et al.* 1987). Because

$NO_3^-$ is rapidly utilised by colonic bacteria (Archer *et al.* 1981), it is possible that the large gut acts as a sink for this metabolite.

Dissimilatory $NO_3^-$ reduction by gut bacteria has been shown to strongly influence the course of some fermentation reactions (Allison and Macfarlane 1988). These workers demonstrated that in mixed populations of colonic bacteria, $H_2$ was the preferred electron donor for $NO_3^-$ reduction and that this inhibited methanogenesis in two ways: Firstly, $NO_3^-$ reducing bacteria outcompeted methanogenic bacteria (MB) for $H_2$ and secondly, the formation of $NO_2^-$ from $NO_3^-$ was directly toxic to MB.

## Methane Production from $H_2$

Approximately 30% - 50% of persons living in Western countries produce $CH_4$ in breath or flatus (Bond *et al.* 1971). Methanogenic bacteria that utilise $H_2$ as an energy source are able to combine it with $CO_2$ according to the equation:

$$4H_2 + CO_2 \text{ ------> } CH_4 + 2H_2O$$

Human colonic methanogens have an obligate requirement for $H_2$, and methanogenesis is an effective and safe method of $H_2$ disposal which occurs in a variety of animals such as the rat, horse, sheep, cow, pig, goose, turkey and chicken (Miller *et al.* 1986). The principal methanogenic species in humans is *Methanobrevibacter smithii,* a Gram positive coccobacillus. This bacterium represents between 0.001-12.6% of the total anaerobe count in faeces, with numbers ranging from about $10^8$ to $10^{10}$ / gram dry weight (Miller and Wolin 1982; 1986). The Gram positive coccus *Methanosphaera stadtmaniae* can also be isolated from faeces in some individuals, but occurs in lower numbers than *M. smithii*, and carries out the following reaction (Miller and Wolin 1983, 1985, 1986):

$$H_2 + CH_3OH \text{ ------> } CH_4 + H_2O$$

Thus, both species have an obligate growth requirement for $H_2$ and must therefore rely on the activities of carbohydrate and amino acid fermenting bacteria in the ecosystem to provide their sources of carbon and energy. Little is known regarding the distribution of MB in the colon, however, we have shown using gut contents from human sudden death victims, that their numbers and activities are higher in the distal than in the proximal colon (Macfarlane *et al.* 1992a). This observation may be related to the difference in pH in different parts of the bowel, since as stated earlier, the proximal colon, where most carbohydrate fermentation occurs, is a low pH region, whereas in the left colon, pH is usually much higher (Cummings *et al.* 1987). *In vitro* fermentation studies support this hypothesis because they have shown that colonic methanogens prefer a neutral to slightly alkaline environment and are inhibited by acidic conditions typical of gut contents in the caecum and ascending colon (Gibson *et al.* 1990c).

In the rumen, methanogenesis may have a profound effect on the end products of carbohydrate fermentation. Hydrogen removal allows re-oxidation of reduced pyridine nucleotides and consequently, increased production of more oxidised SCFA such as acetate and a decrease in propionate formation (Wolin 1974). In man however, little significant difference is evident in SCFA molar ratios in methanogenic and non-methanogenic faecal samples (Wolin and Miller 1983) or in gut contents (Macfarlane and Cummings 1991). This suggests that in this ecosystem, either $H_2$ producing bacteria do not produce much $H_2$ via reduced pyrimidine nucleotides, due to the high levels of gas already present in the colon, or that there is operation of alternative $H_2$ consuming reactions.

**Table 2.7.** Sulphate reducing and methanogenic activities in faecal samples from British (n = 20) and Rural Black South African persons (n = 20)

| Population | Breath methane[a] | SRB count[b] | Sulphate reduction[c] | Breath methane[a] | SRB count[b] | Sulphate reduction[c] |
|---|---|---|---|---|---|---|
| British | 2.9 - 47.2 (6) | ND | 0.01 - 0.06 (6) | ND | 6.7 - 10.2 (14) | 7.6 - 81 (14) |
| Black South African | 3.2 - 50.8 (17) | 5.3 - 6.8 (4) | 0.01 - 0.08 (17) | ND | 8.8 - 10.8 (3) | 3.2 - 264 (3) |

a = ppm; b = $Log_{10}$(g dry wt. faeces)$^{-1}$; c = nmol sulphate reduced $h^{-1}$ (g dry wt. faeces)$^{-1}$
ND = Not detected; The number of positive cases are shown in parenthesis
Taken from Gibson *et al.* (1990a)

## Dissimilatory Sulphate Reduction

One such pathway is through dissimilatory reduction of sulphate by sulphate-reducing bacteria (SRB). It has now been demonstrated by a number of workers that SRB can be present in large numbers in human faeces (Beerens and Romond 1977; Gibson *et al.* 1988a; Lemann *et al.* 1990; Christl *et al.* 1992a). Table 2.7 shows the results of a study on MB and SRB in two geographically diverse human populations. Breath $CH_4$ concentrations, SRB activities and SRB carriage rates were determined in 20 persons living in Cambridge (United Kingdom) and 20 in a rural South African village. It can be seen that the rural Africans were predominantly methanogenic, however a strong inverse relationship between methane production on one hand, and sulphate reduction on the other, was found in both populations. Six of the British subjects and 17 of the Africans were methanogenic. These persons had undetectable, or low SRB counts and negligible sulphate reducing activity. In contrast, the 14 non-methanogenic British and three non-methanogenic South Africans had high sulphate reducer populations (up to 1% of the total microflora) and sulphate reduction rates. These results provide an interesting example of how different types of bacteria can predominate in different human populations, which may possibly be linked to environmental and dietary factors. In general, individuals who harbour SRB in their colons have higher levels of sulphide in their faeces (Fig. 2.11). SRB utilise $H_2$ according to the following equation:

$$4H_2 + SO_4^{2-} + 2H^+ \text{ ------> } H_2S + 4H_2O$$

As with methanogenesis, it is possible to reduce 4 moles of $H_2$ to produce 1 mole of product, which in SRB metabolism is $H_2S$, a highly toxic compound that could have potentially damaging effects on the colonic epithelium (Karrer 1960). As with MB, substrates used by SRB for growth are fermentation products formed by other gut bacteria. All colonic SRB are Gram negative strict anaerobes that can fill the niche occupied by terminal oxidative microorganisms. Many have an obligate requirement for sulphate which is used as an electron acceptor during oxidative reactions, although a restricted number of species are able to grow fermentatively in the absence of $SO_4^{2-}$ (Postgate 1984). So far, species belonging to five different genera of SRB have been

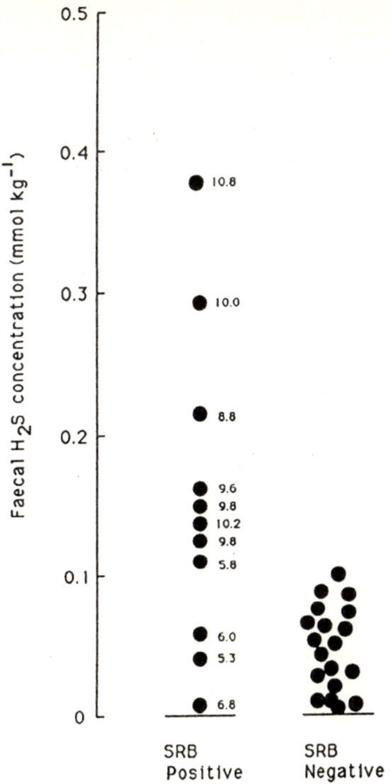


**Fig. 2.11.** Relationship of faecal sulphide concentrations to carriage rates of sulphate-reducing bacteria in methanogenic (SRB negative) and non-methanogenic (SRB positive) individuals. Results are from Gibson *et al.* (1990c).

**Table 2.8.** Substrates utilised by sulphate-reducing bacteria (SRB) isolated from the large intestine

| Substrate | SRB Genus | | | | |
|---|---|---|---|---|---|
| | *Desulfovibrio* | *Desulfobacter* | *Desulfomonas* | *Desulfobulbus* | *Desulfotomaculum* |
| Acetate | - | + | - | - | + |
| Propionate | - | - | - | + | - |
| Butyrate | - | - | - | - | + |
| Lactate | + | - | + | + | - |
| Succinate | + | - | - | - | - |
| Valerate | + | - | - | - | + |
| Glu/Ser/Ala | + | - | - | - | - |
| $H_2/CO_2$ | + | - | - | + | - |
| Ethanol | + | + | + | + | + |
| Pyruvate | + | - | + | + | - |
| Propanol | - | - | - | + | - |
| Butanol | - | - | - | - | + |

+ = growth on substrate; - = no growth on substrate; Glu/Ser/Ala = glutamate/serine/alanine

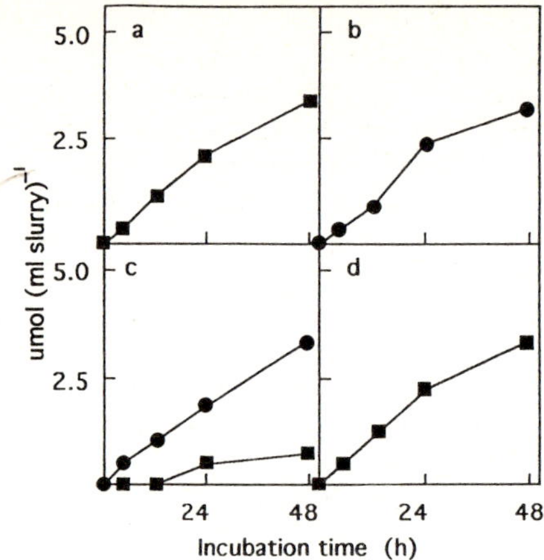


**Fig. 2.12.** Competition for hydrogen between methanogenic bacteria and sulphate-reducing bacteria in faecal slurries. **a**, methanogenic slurries; **b**, non-methanogenic slurries; **c**, mixed slurries; **d**, mixed slurries + 20mM sodium molybdate added. (●) $H_2S$; (■) $CH_4$.

found in the large gut. Their major growth substrates are shown in Table 2.8 where it can be seen that species belonging to the genera *Desulfovibrio* and *Desulfobulbus* can utilise $H_2$ as an electron donor (Gibson *et al.* 1988a). Desulfovibrios are highly motile curved rods, and because they have an incomplete TCA cycle, carry out incomplete oxidation of their substrates, to produce acetate and $CO_2$ (Postgate and Campbell 1966). The genus *Desulfotomaculum* consists of spore forming rod shaped organisms, and these bacteria, with the exception of *Dsm. acetoxidans,* have a restricted metabolic capability similar to desulfovibrios (Campbell and Postgate 1965; Widdel and Pfennig 1977, 1981a). *Desulfomonas* spp. are non-motile straight rods which incompletely oxidise substrates such as pyruvate and ethanol (Moore *et al.* 1976). Non sporing coccobacilli which are able to utilise acetate constitute the genus *Desulfobacter* (Widdel and Pfennig 1981b), whilst *Desulfobulbus* spp. are ellipsoidal or lemon shaped bacteria (Widdel and Pfennig 1982). The numerically predominant genus is *Desulfovibrio*, which constitutes approximately 66% of the total SRB count in gut contents (Gibson *et al.* 1990).

One explanation for the differences found in the distribution of MB and SRB in different individuals noted previously may lie in their relative abilities to compete for $H_2$. Fermentation experiments in our laboratory have shown that colonic SRB can outcompete MB for $H_2$ (Gibson *et al.* 1988b). Fig. 2.12 shows gas and $H_2S$ production data during a 48 h incubation of faecal slurries from three methanogenic and three non-methanogenic volunteers. In the methanogenic individuals, $H_2$ did not accumulate over the time course, but there was significant $CH_4$ production (Fig. 2.12a). In the non-methanogenic slurries however, $H_2S$ was a major product (Fig. 2.12b), confirming active sulphate reduction in these samples. Importantly, when the two types of slurries were mixed in equal proportions, methanogenesis was inhibited, whilst $H_2S$ production was largely unaffected (Fig. 2.12c), indicating efficient competitive utilisation of $H_2$ by SRB at the expense of MB. Methanogenesis only became significant in the mixed samples, when sulphate reduction was inhibited by the addition of 20 mM sodium molybdate (Fig. 2.12d). The physiological explanation for these results lies in the greater affinity of SRB for $H_2$, when compared to MB (Ks of *Desulfovibrio vulgaris*, 1mmol $l^{-1}$; Ks of *Methanobrevibacter smithii*, 6 mmol $l^{-1}$(Kristjansson *et al.* 1982). Furthermore, the oxidation of $H_2$ by SRB is thermodynamically more favourable ($\Delta G^{o'} = -152.2$ kJ per mol) than by MB ($\Delta G^{o'} = -131$ kJ per mol) (Thauer *et al.* 1977), which may also indirectly influence the outcome of competition.

In other natural environments, competition for $H_2$ has been shown to be largely dependent on sulphate availability, with SRB outcompeting MB in the presence of sulphate (Winfrey and Zeikus 1977; Oremland and Taylor 1978; Mountford *et al.* 1980; Robinson and Tiejde 1984). A similar situation seems to occur in man as evidenced by the fact that the addition of 15mmol $d^{-1}$ sulphate to the diet of six methanogenic volunteers could result in competitive displacement of MB within a few days and the appearance of hitherto undetected SRB in their faeces (Christl *et al.* 1992b). This effect was found in only half of the volunteers however, suggesting that the two types of bacteria co-exist in some but not all individuals.

The sulphate content of a normal Western diet is variable and ranges from 2-16 mmol $d^{-1}$. Maximum absorption in the small bowel occurs at levels below 7 mmol $d^{-1}$, however, at values above this, there is increasing loss to the large gut (Florin *et al.* 1991). Foods with a high sulphate content include dried fruits, fermented beverages and white bread (Florin and Cummings - unpublished results) and it is possible that individual dietary preferences and daily variations in the diet might influence competition between MB and SRB. There is also evidence that endogenous sources of sulphate are of some importance. This is because a variety of sulphate containing substances are produced by the host, such as the highly sulphated polysaccharide mucin which is constantly secreted by goblet cells in the colonic epithelium, or chondroitin sulphate which enters the colon in meat residues and sloughed epithelial cells from the gastrointestinal tract (Vercellotti *et al.* 1977; Salyers and O'Brien 1980). Both substances are extensively degraded by the gut microflora (Gibson *et al.* 1988b) and may release free sulphate (Tsai *et al.* 1992), which will become available for SRB. Batch culture studies with mixed populations of gut bacteria have shown elevated $H_2S$ production from both mucin and chondroitin sulphate when compared to starch, and that when both MB and SRB are present in mixed culture, sulphate reduction was preferentially stimulated by mucin addition (Gibson *et al.* 1988b,c).

## Acetogenesis

The reduction of $CO_2$ by $H_2$; as carried out by acetogenic bacteria (AB) occurs as shown below:

$$4H_2 + 2CO_2 \text{ ------> } CH_3COO^- + H^+ + 2H_2O$$

This reaction has a $\Delta G^{o'}$ of -95kJ per mol making it considerably less efficient energetically than either methanogenesis or dissimilatory sulphate reduction (Thauer *et al.* 1977). Nevertheless, many strictly anaerobic bacteria are able to carry out acetogenesis (for review, see Fuchs 1986), some of which, such as *Eubacterium limosum* (Genthner and Bryant 1982) and *Peptostreptococcus productus* (Lorowitz and Bryant 1984) are inhabitants of the human colon. However, the fact that these bacteria can be shown to carry out acetogenic reactions under certain conditions *in vitro* does not necessarily mean that they also do so during growth in the gut. This is because both species are saccharolytic and under normal circumstances in the gut may prefer to ferment carbohydrate, which is an energetically more advantageous process than growing on $H_2$ and $CO_2$. The individual substrate preferences of putatively acetogenic bacteria is therefore an important problem that needs to be addressed. A good example of this comes from the work of Greening and Leedle (1989) who isolated acetogenic bacteria from the bovine rumen (a predominantly methanogenic ecosystem) that can also grow on carbohydrates.

The importance of acetogenesis in the human large bowel therefore remains to be established. Studies by Gibson *et al.* (1990c) indicated that low levels of $H_2$-dependent $CO_2$ reduction to acetate occurred during *in vitro* incubations of human faecal bacteria, and that either methanogenesis or sulphate reduction normally predominated. However, different results were obtained by Lajoie *et al.* (1988) who found significant levels of acetogenesis during incubations of faecal bacteria, but only in the absence of methanogenesis, suggesting that there was competition between AB and MB for $H_2$.

It is possible that the different results obtained in these two studies were influenced by the fact that different types of individual were being looked at, since one group of subjects was in the United Kingdom, and the other was in the United States. However, there were also methodological differences in the experimental approaches adopted by both laboratories, in that one group followed the incorporation of $^{14}$C-labelled $HCO_3^-$ into acetate and depended on endogenous $H_2$ generation in their fermentation systems for acetate production, whereas the other group used a $^{13}$C-labelled substrate with nuclear magnetic resonance techniques, and incubated the cell suspensions under an atmosphere containing 80% $H_2$.

Comparisons between humans and animals would appear to be of little value in determining the significance of acetogenesis in man, because the relative importance of this process in overall gut metabolism varies markedly between different species. For example, in the termite gut, the reduction of $CO_2$ with $H_2$ contributes about one third of the total acetate produced (Breznak and Switzer 1986), whereas in some rodents, acetogenesis occurs, but only in the absence of significant levels of methanogenesis (Prins and Lankhorst 1977), suggesting that there is competition for $H_2$ between MB and AB in these animals. In the pig however, $CH_4$ formation and acetogenesis were found to occur simultaneously (DeGraeve *et al.* 1990).

## References

Allison C, Macfarlane GT (1988) Effect of nitrate on methane production by slurries of human faecal bacteria. J Gen Microbiol 134: 1397-1405

Allison C, Macfarlane GT (1989a) Protease production by *Clostridium perfringens* in batch and continuous culture. Lett Appl Microbiol 9: 45-48

Allison C, Macfarlane GT (1989b) Influence of pH, nutrient availability and growth rate on amine production by *Bacteroides fragilis* and *Clostridium perfringens*. Appl Environ Microbiol 55: 2894-2898

Allison C, Macfarlane GT (1989c) Dissimilatory nitrate reduction by *Propionibacterium acnes*. Appl Environ Microbiol 55: 2889-2903

Allison C, Macfarlane GT (1990) Regulation of protease production in *Clostridium sporogenes*. Appl Environ Microbiol 56: 3485-3490

Andreesen JR, Bahl H, Gottschalk G (1989) Introduction to the physiology and biochemistry of the genus *Clostridium*. In: Minton NP, Clarke DJ (eds) Clostridia, biotechnology handbooks 3. Plenum Press, New York pp 27-62

Anon (1968) Headache, tyramine, serotonin and migraine. Nutr Rev 26: 40-44

Aragozzini F, Ferrari A, Pacini N, Sualandris R (1979) Indole-3-lactic acid as a tryptophan metabolite produced by *Bifidobacterium* spp. Appl Environ Microbiol 38: 544-546

Babu S, Chandler H, Batish UK, Bhatia KL (1986) Factors affecting amine production in *Streptococcus cremoris*. Food Microbiol 3:359-362

Bakke OM (1969) Studies on the degradation of tyrosine by rat caecal contents. Scan J Gastroenterol 4: 603-608

Banwell JG, Branch WJ, Cummings JH (1981) The microbial mass in the human large intestine. Gastroenterology 80: 1104

Barker HA (1981) Amino acid degradation by anaerobic bacteria. Ann Rev Biochem 50: 23-40

Bartholomew B, Hill MJ (1984) The pharmacology of dietary nitrate and the origin of urinary nitrate. Food Chem Toxicol 22: 789-795

Beeken W, Northwood I, Beliveau C, Gump D (1987) Phagocytes in cell suspensions of human colon mucosa. Gut 28: 976-980

Beerens H, Romond C (1977) Sulfate-reducing anaerobic bacteria in human feces. Am J Clin Nutr 30: 1770-1776

Berg JO (1981) Cellular location of glycoside hydrolases in *Bacteroides fragilis*. Curr Microbiol 5: 13-17

Bezkorovainy A, Miller-Catchpole R (1989) Biochemistry and physiology of bifidobacteria. CRC Press, Boca Raton

Bingham SA, Pett S, Day KC (1990) NSP intake of a representative sample of British adults. J Hum Nutr Diet 3: 339-344

Blackwood AC, Neish AC, Ledingham GA (1956) Dissimilation of glucose at controlled pH values by pigmented and non-pigmented strains of *Escherichia coli*. J Bacteriol 72: 497-499

Boeker EA, Snell EE (1972) Amino acid decarboxylases. The enzymes 6: 217-253

Bohe M, Borgstrom S, Genell S, Ohlsson GK (1983) Determination of immunoreactive trypsin, pancreatic elastase and chymotrypsin in extracts of human faeces and ileostomy drainage. Digestion 27: 8-15

Bond JH, Currier BE, Buchwald H, Levitt MD (1980) Colonic conservation of malabsorbed carbohydrate. Gastroenterology 78: 444-447

Bond JH, Engel RR, Levitt MD (1971) Factors influencing pulmonary methane excretion in man. J Exp Med 133: 572-578

Bornside GH (1978) Stability of human fecal flora. Am J Clin Nutr 31: 5141-5144

Botsford JL, Demoss RD (1972) *Escherichia coli* tryptophanase in the enteric environment. J Bacteriol 109: 74-80

Boulton AA, Cokkson B, Paulton R (1970) Hypertensive crisis in a patient on MAOI antidepressants following a meal of beef liver. Can Med Ass J 102: 1394-1395

Breznak JA, Pankatz HS (1977) In situ morphology of the gut microbiota of wood eating termites [*Reticulitennes flaviceps* Kollar and *Coptotermes formosanus* Shiraki]. Appl Environ Microbiol 33: 406-426

Breznak JA, Switzer JM (1986) Acetate synthesis from $H_2$ plus $CO_2$ by termite gut microbes. Appl Environ Microbiol 52: 623-630

Brock TD (1966) Principles of microbial ecology. Prentice Hall, New Jersey

Brodribb AJM, Groves C (1978) Effect of bran particle size on stool weight. Gut 19: 60-63

Bryan GT (1971) The role of urinary tryptophan metabolites in the etiology of bladder cancer. Am J Clin Nutr 24: 841-847

Burnham WR, Lennard-Jones JE (1978) Mycobacteria as a possible cause of inflammatory bowel disease. Lancet 693-696

Calloway DH, Murphy EL (1968) The use of expired air to measure intestinal gas formation. Ann NY Acad Sci 150: 82-95

Campbell LL, Postgate JR (1965) Classification of the spore forming sulfate-reducing bacteria. Bacteriol Rev 29: 359-363

Carlson JR, Yokohama MT, Dickenson EO (1972) Induction of pulmonary edema and emphysema in cattle and goats with 3-methylindole. Science 176: 298-299

Chacko A, Cummings JH (1988) Nitrogen losses from the human small bowel: obligatory losses and the effect of physical form of food. Gut 29: 809-815

Christensen J (1991) Gross and microscopic anatomy of the large intestine. In: Phillips SF, Pemberton JH, Shorter RG (eds) The large intestine: physiology, pathophysiology and disease. Raven Press Ltd, New York pp 13-35

Christl SU, Gibson GR, Cummings JH (1992b) The role of dietary sulphate in the regulation of methanogenesis in the human large intestine. Gut 33 - in press

Christl SU, Murgatroyd PR, Gibson GR, Cummings JH (1990) Quantitative measurement of hydrogen and methane from fermentation using a whole body calorimeter. Gastroenterology 98: A164

Christl SU, Murgatroyd PR, Gibson GR, Cummings JH (1992a) Production, metabolism, and excretion of hydrogen in the large intestine. Gastroenterology 102: 1269-1277

Conrad R, Phelps TJ, Zeikus JG (1985) Gas metabolism evidence in support of the juxtaposition of hydrogen-producing and methanogenic bacteria in sewage sludge and lake sediments. Appl Environ Microbiol 50: 595-601

Croucher SC, Houston AP, Bayliss CE, Turner RJ (1983) Bacterial populations associated with different regions of the human colon wall Appl Environ Microbiol 45, 1025-1033

Cummings JH (1978) Diet and transit through the gut. J Plant Foods 3: 83-95

Cummings JH (1981) Short chain fatty acids in the human colon. Gut 22: 763-779

Cummings JH (1983) Dietary fibre and the intestinal microflora. In: Hallgren B (ed) Nutrition and the intestinal flora. Swedish Nutrition Foundation, Stockholm pp 77-86

Cummings JH (1987) Dietary fiber. Am J Clin Nutr 45: 1040-1043

Cummings JH, Banwell JG, Englyst HN, Coleman N, Segal I, Bersohn D (1990) The amount and composition of large bowel contents. Gastroenterology 98: A408

Cummings JH, Bingham SA, Heaton KW, Eastwood MA (1993) Fecal weight, colon cancer risk and dietary intake of non-starch polysaccharides (dietary fiber). Gastroenterology 103: 1783-1789

Cummings JH, Branch WJ (1986) Fermentation and the production of short-chain fatty acids in the human large intestine. In: Vahormy GV, Kritchevsky, D (eds) Dietary fibre. Plenum Press, New York pp 131-149

Cummings JH, Gibson GR, Macfarlane GT (1989) Quantitative estimates of fermentation in the hind gut of man. Acta Veterin Scan 86: 76-82

Cummings JH, Hill MJ, Bone ES, Branch WJ, Jenkins DJA (1979) The effect of meat protein and dietary fiber on colonic function and metabolism. Part II Bacterial metabolites in feces and urine. Am J Clin Nutr 32: 2094-2101

Cummings JH, Pomare EW, Branch WJ, Naylor CPE, Macfarlane GT (1987) Short chain fatty acids in human large intestine, portal, hepatic and venous blood. Gut 28: 1221-1227

Dalgliesh CE, Kelley W, Horning EC (1958) Excretion of a sulphatoxyl derivative of skatole in pathological states in man. Biochem J 70: 13P

Degnan BA, Macfarlane GT (1991) Comparison of carbohydrate substrate preferences in eight species of bifidobacteria. FEMS Microbiol Letts 84: 151-156

DeGraeve KG, Grivet JP, Durand M, Beaumartin P, Demeyer D (1990) NMR studies of $^{13}CO_2$ incorporation into short-chain fatty acids by pig large-intestinal flora. Can J Microbiol 36: 579-582

Demoss RD, Moser K (1969) Tryptophanase in diverse bacterial species. J Bacteriol 98: 1267

DeQuattro VL, Sjoerdsma A (1967) Origin of urinary tyramine and tryptamine. Clin Chem Acta 16: 227-233

Duerden BI (1980) The isolation and identification of *Bacteroides* spp from the normal human faecal flora. J Med Microbiol 13: 69-78

El-Soda M, Bergere JL, Desmazeard MJ (1978) Detection and localisation of peptide hydrolases in *Lactobacillus casei*. J Dairy Res 45: 519-524

Elsden SR, Hilton MD (1978) Volatile fatty acid production from threonine, valine, leucine and isoleucine by clostridia. Arch Microbiol 117: 165-172

Elsden SR, Hilton MG, Waller JM (1976 ) The end products of the metabolism of aromatic amino acids by clostridia. Arch Microbiol 107: 283-288

Englyst HN, Bingham SA, Runswick SA, Collinson E, Cummings JH (1988) Dietary fibre (non starch polysaccharides) in fruit, vegetables and nuts. J Hum Nutr Res 1: 247-286

Englyst HN, Bingham SA, Runswick SA, Collinson E, Cummings JH (1989) Dietary fibre (non-starch polysaccharides) in cereal products. J Human Nutr and Diet 2: 253-271

Englyst HN, Cummings JH (1987) Resistant starch, a 'new' food component: A classification of starch for nutritional purposes. In: Marton ID (ed) Cereals in a European context. Ellis Horwood, Chichester pp 221-233

Englyst HN, Hay S, Macfarlane GT (1987) Polysaccharide breakdown by mixed populations of human faecal bacteria. FEMS Microbiol Ecol 95: 163-171

Englyst HN, Macfarlane GT (1986) Breakdown of resistant and readily digestible starch by human gut bacteria. J Sci Food Agri 37: 699-706

Etterlin C, McKeown A, Bingham SA, Elia M, Macfarlane GT, Cummings JH (1992) D-Lactate and acetate as markers of fermentation in man. Gastroenterology 102: A551

Fellers CR, Clough RW (1925) Indole and skatole determination in bacterial cultures. J Bacteriol 10: 105-133

Finegold SM, Sutter VL (1978) Fecal flora in different populations, with special reference to diet. Am J Clin Nutr 31: 51116-5122

Finegold SM, Attlebury HR, Sutter VL (1974) Effect of diet on human fecal flora: comparison of Japanese and American diets. Am J Clin Nutr 27: 1456-1469

Finegold SM, Sutter VL, Mathisen GE (1983) Normal indigenous intestinal flora. In: Hentges DJ (ed) Human intestinal microflora in health and disease. Academic Press, London pp 3-31

Florin THJ, Neale G, Gibson GR, Christl SU, Cummings JH (1991) Metabolism of dietary sulphate: absorption and excretion in humans. Gut 32: 766-773

Freter R (1983) Mechanisms that control the microflora. In: Hentges DJ (ed) Human intestinal microflora in health and disease. Academic Press, London pp 35-54

Fuchs G (1986) $CO_2$ fixation in acetogenic bacteria: variations on a theme. FEMS Microbiol Rev 39: 181-213

Gale EF (1946) The bacterial amino acid decarboxylases. Adv Enzymol 6: 1-32

Gale EF, Epps HMR (1942) The effect of the pH of the medium during growth on the enzymic activity of bacteria (*Escherichia coli* and *Micrococcus lysodeikticus*) and the biological significance of the changes produced. Biochem J 36: 600-618

Genthner BRS, Bryant MP (1982) Growth of *Eubacterium limosum* with carbon monoxide as the energy source. Appl Environ Microbiol 43: 70-74

Gibson GR (1990) A review: Physiology and ecology of the sulphate-reducing bacteria. J Appl Bacteriol 69: 76 9-797

Gibson GR, Macfarlane GT, Cummings JH (1988a) Occurrence of sulphate-reducing bacteria in human faeces and the relationship of dissimilatory sulphate reduction to methanogenesis in the large gut. J Appl Bacteriol 65: 103-111

Gibson GR, Cummings JH, Macfarlane GT (1988b) Competition for hydrogen between sulphate-reducing bacteria and methanogenic bacteria from the human large intestine. J Appl Bacteriol 65: 241-247

Gibson GR, Cummings JH, Macfarlane GT (1988c) Use of a three-stage continuous culture system to study the effect of mucin on dissimilatory sulfate reduction and methanogenesis by mixed populations of human gut bacteria. Appl Environ Microbiol 54: 2750-2755

Gibson GR, Cummings JH, Macfarlane GT (1990a) Factors affecting hydrogen uptake by bacteria growing in the human large intestine. In: Belaich JP, Bruschi M, Garcia JL (eds) Microbiology and biochemistry of strict anaerobes involved in interspecies hydrogen transfer. Plenum Press, New York pp 191-201

Gibson GR, Macfarlane S, Cummings JH (1990b) The fermentability of polysaccharides by mixed human faecal bacteria in relation to their suitability as bulk-forming laxatives. Lett Appl Microbiol 11: 251-254

Gibson GR, Cummings JH, Macfarlane GT, Allison C, Segal I, Vorster HH, Walker ARP (1990c) Alternative pathways for hydrogen disposal during fermentation in the human colon. Gut 31: 679-683

Gibson JA, Sladen GE, Dawson AM (1976) Protein absorption and ammonia production: the effects of dietary protein and removal of the colon. Brit J Nutr 35: 61-65

Gibson SAW, Macfarlane GT (1988) Studies on the proteolytic activity of *Bacteroides fragilis*. J Gen Microbiol 134: 19-27

Gibson SAW, Mcfarlan C, Hay S, Macfarlane GT (1989) Significance of microflora in proteolysis in the colon. Appl Environ Microbiol 55: 679-683

Gorbach SL (1967) Population control in the small bowel. Gut 8: 530-532

Gorbach SL, Nahas L, Lerner PI, Weinstein L (1967) Studies of intestinal microflora. I Effects of diet, age and periodic sampling on numbers of fecal microorganisms in man. Gastroenterology 53: 845-855

Greening RC, Leedle JAZ (1989) Enrichment and isolation of *Acetitomaculum ruminis*, gen nov, sp nov: acetogenic bacteria from the bovine rumen. Arch Microbiol 151: 399-406

Hausmann E, Kaufman E (1969) Collagenase activity in a particulate fraction of *Bacteroides melaninogenicus*. Biochim Biophys Acta 194: 612-615

Heller SN, Hackler LR, Rivers JM (1980) Dietary fibre: the effect of particle size of wheat bran on colonic function in young adult men. Am J Clin Nutr 33: 1734-1744

Hespell RB, Bryant MP (1979) Efficiency of rumen microbial growth: Influence of some theoretical and experimental factors on $Y_{ATP}$. J Anim Sci 49: 1640-1658

Holdeman LV, Cato EP, Moore WEC (1977) Anaerobe laboratory manual, 4th edn Virginia Polytechnic Institute Anaerobic Laboratory, Blacksburg

Holdeman LV, Good IJ, Moore WEC (1976) Human fecal flora: variation in bacterial composition between individuals and a possible effect of emotional stress. Appl Environ Microbiol 31:359-375

Horning EC, Dalgleish DE (1958) The association of skatole forming bacteria in the small intestine with the malabsorption syndrome and certain anaemias. Biochem J 70: 13-14P

Hoskins LC, Boulding ET (1981) Mucin degradation in human colonic ecosystems. J Clin Invest 67: 163-172

Hosoya N, Dhorraninstra B, Hidaka H (1988) Utilisation of [U-$^{14}$C] fructo-oligosaccharides in man as energy resources. J Clin Biochem Nutr 5: 67-74

Hummel FC, Shepherd ML, Macy IG (1943) Disappearance of cellulose and hemicellulose from the digestive tract of children. J Nutr 25: 59-70

Illman RJ, Topping DL, McIntosh GH, Trimble RP, Stoner GB, Taylor MN, Cheng BG (1988) The hypercholesterolaemic effects of dietary propionate: studies in whole animals and perfused rat liver. Ann Nutr Met 32: 97-107

Ingram E, Holland KT, Gowland G, Cunliffe WJ (1983) Studies of the extracellular proteolytic activity produced by *Propionibacterium acnes*. J Appl Bacteriol 54: 263-271

Irvine WT, Buthie HL, Watson WG (1959) Urinary output of free histamine after a meal. Lancet 1061-1064

Jeanloz RW (1970) Mucopolysaccharides of higher animals. In: Pigman W, Horton D, Herp A (eds) The carbohydrates. Academic Press, New York pp 589-625

Johnson KA (1977) The production of secondary amines by human gut bacteria and its possible relevance to carcinogenesis. Med Lab Sci 34: 131-143

Jones WJ, Nasgle DP Jnr, Whitman WB (1987) Methanogens and the diversity of archaebacteria. Microb Rev 56: 135-177

Juhlen-Dannfelt A (1977) Ethanol effects of substrate utilisation by the human brain. Scan J Clin Lab Invest 37: 443-449

Karrer P (1960) Organic chemistry, Elsevier Press, New York

Keith SM, Macfarlane GT, Herbert RA (1982) Dissimilatory nitrate reduction by a strain of *Clostridium butyricum* isolated from estuarine sediments. Arch Microbiol 132: 62-66

Knight TM, Forman D, Al-Dabbagh SA, Doll R (1987) Estimation of dietary intake of nitrate and nitrite in Great Britain. Food Chem and Toxicol 25: 277-285

Kristjansson JK, Shonheit P, Thauer RK (1982) Different Ks values for hydrogen of methanogenic bacteria and sulfate reducing bacteria: an explanation for the apparent inhibition of methanogenesis by sulfate. Arch Microbiol 131: 278-282

Kuknal J, Adams A, Preston F (1965) Protein producing capacity of the human exocrine pancreas. Surgery 162: 67-73

Lajoie SF, Bank S, Miller TL, Wolin MJ (1988) Acetate production from hydrogen and [$^{13}$C] carbon dioxide by the microflora of human feces. Appl Environ Microbiol 54: 2723-2727

Launiala K (1968a) The effect of unabsorbed sucrose or mannitol. Scan J Gastroenterol 39: 665-671

Launiala K (1968b) The mechanisms of diarrhoea in congenital disaccharide malabsorption. Acta Paed Scan 57: 425-432

Lee A (1980) Normal flora of animal intestinal surfaces. In: Bitton G, Marshall KC (eds) Adsorption of microorganisms to surfaces. John Wiley, New York pp 145-174

Lee FD, Kraszewski A, Gordon J, Howie JGR, McSeveney D, Harland WA (1971) Intestinal spirochaetosis. Gut 12: 126-133

Lee K, Greger JL, Consaul MS, Chinn MS (1986) Nitrate, nitrite balance, and de novo synthesis of nitrate in humans consuming cured meats. Am J Clin Nutr 44: 188-194

Lemann F, Pochart P, Pellier P, Flourie B, Rambaud JC (1990) Does sulfate availability in colonic content control methanogenesis in man? Gastroenterology 98: A186

Levitt MD (1971) Volume and composition of human intestinal gas determined by means of an intestinal wash out system. New Eng J Med 284: 1394-1398

Lindeneg O, Mellemgaard K, Fabricus J, Lundqvist F (1964) Myocardial utilisation of acetate, lactate and free fatty acids after ingestion of ethanol. Clin Sci 27: 427-435

Lorowitz WH, Bryant MP (1984) *Peptostreptococcus productus* strain that grows rapidly with $CO_2$ as the energy source. Appl Environ Microbiol 47: 961-964

Lundqvist F, Sestoft L, Damgaard SE, Clavsen JP , Trap-Jensen J (1973) Utilisation of acetate in the human forearm during exercise after ethanol ingestion. J Clin Invest 52: 3231-3235

Macfarlane GT, Allison C (1986) Utilisation of protein by human gut bacteria. FEMS Microbiol Ecol 38: 19-24

Macfarlane GT, Cummings JH (1991) The colonic flora, fermentation, and large bowel digestive function. In: Phillips SF, Pemberton JH, Shorter RG (eds) The large intestine: physiology, pathophysiology and disease. Raven Press Ltd, New York pp 51-92

Macfarlane GT, Englyst HN (1986) Starch utilisation by the human large intestinal microflora. J Appl Bacteriol 60: 195-201

Macfarlane GT, Gibson GR (1991a) Formation of glycoprotein degrading enzymes by *Bacteroides fragilis*. FEMS Microbiol Lett 77: 289-294

Macfarlane GT, Gibson GR (1991b) Co-utilization of polymerized carbon sources by *Bacteroides ovatus* grown in a two-stage continuous culture system. Appl Environ Microbiol 57: 1-6

Macfarlane GT, Macfarlane S (1991) Utilization of pancreatic trypsin and chymotrypsin by proteolytic and non proteolytic *Bacteroides fragilis* type bacteria. Curr Microbiol 23: 143-148

Macfarlane GT, Macfarlane S (1992) Physiological and nutritional factors affecting synthesis of extracellular metalloproteases by *Clostridium bifermentans* NCTC 2914. Appl Environ Microbiol 58: 1195-1200

Macfarlane GT, Allison C, Gibson GR (1988a) Effect of pH on protease activities in the large intestine. Lett Appl Microbiol 7: 161-164

Macfarlane GT, Allison C, Gibson SAW, Cummings JH (1988b) Contribution of the microflora to proteolysis in the human large intestine. J Appl Bacteriol 64: 37-46

Macfarlane GT, Cummings JH, Allison C (1986) Protein degradation by human intestinal bacteria. J Gen Microbiol 132: 1647-1656

Macfarlane GT, Cummings JH, Macfarlane S, Gibson GR (1989) Influence of retention time on degradation of pancreatic enzymes by human colonic bacteria grown in a 3-stage continuous culture system. J Appl Bacteriol 67: 521-527

Macfarlane GT, Gibson GR, Cummings JH (1992a) Comparison of fermentation reactions in different regions of the colon. J Appl Bacteriol 72: 57-64

Macfarlane GT, Gibson GR, Beatty E, Cummings JH (1992c) Estimation of short chain fatty acid production from protein by human intestinal bacteria based on branched-chain fatty acid measurements. FEMS Microbiol Ecol - in press

Macfarlane GT, Hay S, Macfarlane S, Gibson GR (1990) Effect of different carbohydrates on growth, polysaccharidase and glycosidase production by *Bacteroides ovatus*, in batch and continuous culture. J Appl Bacteriol 68: 179-187

Macfarlane GT, Macfarlane S, Gibson GR (1992b) Synthesis and release of proteases by *Bacteroides fragilis*. Curr Microbiol 24: 55-59

Macy JM, Probst I (1979) The biology of gastrointestinal bacteroides. Ann Rev Microbiol 33: 561-594

Meijer-Severs GJ, Van Santen E (1986) Variations in the anaerobic faecal flora of ten healthy human volunteers with special reference to the *Bacteroides fragilis* group and *Clostridium difficile*. Zbl Bakt Hyg 261: 43-52

Metchnikoff E (1905) The nature of man. GP Putnams and Sons, New York

Midvedt T, Norman A (1968) Parameters in 7-a-dehydroxylation of bile acids by anaerobic lactobacilli. Acta Pathol Microbiol Scand 72: 313-329

Miller TL, Wolin MJ (1982) Enumeration of *Methanobrevibacter smithii* in human feces. Arch Microbiol 131: 14-18

Miller TL, Wolin MJ (1983) Oxidation of hydrogen and reduction of methanol to methane is the sole energy source for a methanogen isolated from human feces. J Bacteriol 153: 1051-1055

Miller TL, Wolin MJ (1985) *Methanosphaera stadtmaniae* gen nov, sp nov: a species that forms methane by reducing methanol with hydrogen. Arch Microbiol 141: 116-122

Miller TL, Wolin MJ (1986) Methanogens in human and animal intestinal tracts. System Appl Microbiol 7: 223-229

Miller TL, Weaver GA, Wolin MJ (1984) Methanogens and anaerobes in a colon segment isolated from the normal fecal stream. Appl Environ Microbiol 48: 449-450

Miller TL, Wolin MJ, Kusel EA (1986) Isolation and characterisation of methanogens from animal feces. System Appl Microbiol 8: 234-238

Mitsuoka T (1984) Taxonomy and ecology of bifidobacteria. Bifid Microf 3: 11-28

Moore WEC, Holdeman LV (1974) Human fecal flora. The normal flora of 20 Japanese-Hawaiians. Appl Environ Microbiol 27: 961-979

Moore WEC, Johnson JL, Holdeman LV (1976) Emendation of *Bacteroidaceae* and *Butyrivibrio* and description of *Desulfomonas*,, gen nov and ten new species in the genera *Desulfomonas*, *Butyrivibrio*, *Eubacterium*, *Clostridium* and *Ruminococcus*. Int J Syst Bacteriol 26: 238-252

Morris DR, Boeker EA (1983) Biosynthetic and biodegradative ornithine and arginine decarboxylase from *Escherichia coli*. Meth Enzymol 94: 125-134

Morris DR, Fillingame RH (1974) Regulation of amino acid decarboxylation. Ann Rev Biochem 43: 303-325

Mountford DO, Asher RA, Mays EL, Tiedje JM (1980) Carbon and electron flow in mud and sand intertidal sediments at Delaware Inlet, Nelson, New Zealand. Appl Environ Microbiol 39: 686-694

Murray KE, Adams RS, Earl JW, Shaw KJ (1986) Studies of the free faecal amines of infants with gastroenteritis and of healthy infants. Gut 27: 1173-1180

Nagase M, Matsuo T (1982) Interactions between amino acid degrading bacteria and methanogenic bacteria in anaerobic digestion. Biotech Bioeng 24: 2227-2239

Nelson DP, Mata LJ (1970) Bacterial flora associated with the human gastrointestinal mucosa. Gastroenterology 58: 56-61

Nisman B (1954) The Stickland reaction. Bacteriol Rev 1: 16-42

Oremland RS, Taylor BF (1978) Sulfate reduction and methanogenesis in marine sediments. Geochim Cosmochim Acta 42: 209-214

Pettipher GL, Latham M(1979) Production of enzymes degrading plant cell walls and fermentation of cellobiose by *Ruminococcus flavifaciens*. J Gen Microbiol 110: 29-38

Phear EA, Ruebner B (1956) The *in vitro* production of ammonium and amines by intestinal bacteria in relation to nitrogen toxicity as a factor in hepatic coma. Br J Exp Path 37: 253-262

Pittman KA, Lakshmanan S, Bryant MP (1967) Oligopeptide uptake by *Bacteroides ruminicola*. J Bacteriol 93: 1499-1508

Postgate JR (1984) The sulphate-reducing bacteria, 2nd edn Cambridge University Press, Cambridge

Postgate JR, Campbell LL (1966) Classification of *Desulfovibrio* species, the non-sporulating sulfate-reducing bacteria. Bacteriol Rev 30: 732-738

Prins RA, Lankhorst A (1977) Synthesis of acetate from $CO_2$ in the cecum of some rodents. FEMS Microb Letts 1: 255-258

Rafii F, Franklin W, Cerniglia CE (1990) Azoreductase activity of anaerobic bacteria isolated from human intestinal microflora. Appl Environ Microbiol 56: 2146-2151

Ramakrishna BS, Gee D, Weiss A, Pannall P, Robert-Thomas, IC, Roediger WEW (1989) Estimation of phenolic conjugation by colonic mucosa. J Clin Pathol 42: 620-623

Reddy BS, Weisburger JH, Wynder EL (1975) Effect of high risk and low risk diets for colon carcinogenesis of faecal microflora and steroids of man. J Nutr 105: 878-884

Robinson JA, Tiedje JM (1982) Kinetics of hydrogen consumption by rumen fluid, anaerobic digestor sludge, and sediment. Appl Environ Microbiol 44: 1374-1384

Robinson JA, Tiedje JM (1984) Competition between sulfate-reducing and methanogenic bacteria for hydrogen under resting and growing conditions. Arch Microbiol 137: 26-32

Roediger WEW (1980) Role of anaerobic bacteria in the metabolic welfare of the colonic mucosa of man. Gut 21: 793-798

Rogers P (1986) Genetics and biochemistry of *Clostridium* relevant to development of fermentation processes, Adv Appl Microbiol 31: 1-60

Russell JB, Baldwin RL (1978) Substrate preferences in rumen bacteria: evidence of catabolite regulatory mechanisms. Appl Environ Microbiol 36: 319-329

Salyers AA, Leedle JAZ (1983) Carbohydrate metabolism in the human colon. In: Hentges DJ (ed) Human intestinal microflora in health and disease. Academic Press, London pp 129-146

Salyers AA, O'Brien M (1980) Cellular location of enzymes involved in chondroitin sulfate breakdown by *Bacteroides thetaiotaomicron* J Bacteriol 143: 777-780

Savage DC (1977) Microbial ecology of the gastrointestinal tract. Ann Rev Microbiol 31: 107-133

Scardovi V (1986) Genus *Bifidobacterium*. In: Mair NS (ed) Bergey's manual of systematic bacteriology, vol 2 Williams and Wilkins, New York pp 1418-1434

Schmidt EG (1949) Urinary phenols. Simultaneous determination of phenol and p-cresol in urine. J Biol Chem 179: 211-215

Shamsuddin AM (1989) The large intestinal mucosa. In: Whitehead R (ed) Gastrointestinal and oesophageal pathology. Churchill Livingstone, Edinburgh pp 41-49

Smith AC, Podolsky DK (1986) Colonic mucin glycoproteins in health and disease. In: Mendelkoff A (ed) Clinics in gastroenterology, WB Saunders, Philadelphia pp 815-837

Smith TA (1980) Amines in food. Food Chem 6: 169-200

Southgate DAT, Branch WJ, Hill MJ (1976) Metabolic responses to dietary supplements of bran. Metabolism 25: 1129-1135

Stams AJM, Hansen TA (1984) Fermentation of glutamate and other compounds by *Acidaminobacter hydrogenoformans* gen nov, sp nov, an obligate anaerobe isolated from black mud. Studies with pure cultures and mixed cultures with sulfate-reducing and methanogenic bacteria. Arch Microbiol 137: 329-337

Stephen AM, Cummings JH (1979) The influence of dietary fiber in faecal nitrogen excretion in man. Proc Nutr Soc 38: A141

Stephen AM, Cummings JH (1980) The microbial contribution to human faecal mass. J Med Microbiol 13: 45-56

Stephen AM, Haddad AC, Philips SF (1983) Passage of carbohydrate into the colon. Direct measurements of humans. Gastroenterology 85: 589-595

Stephen AM, Wiggins HS, Cummings JH (1987) Effect of changing transit time on colonic microbial metabolism in man. Gut 28: 610-609

Sumner S, Speckland M, Somers E, Taylor S (1985) Isolations of histamine-producing *Lactobacillus buchneri* from Swiss cheese implicated in food poisoning outbreak. Appl Environ Microbiol 50: 1094-1096

Tadesse K, Smith D, Eastwood MA (1980) Breath hydrogen ($H_2$) and methane ($CH_4$) excretion patterns in normal man and in clinical patients. Quart J Exp Physiol 65: 85-97

Thacker PA, Bowland JP (1981) Effects of dietary propionic acid on serum lipids and lipoproteins of pigs fed diets supplemented with soybean meal or canola meal. Can J An Sci 61: 439-448

Thauer RK, Jungermann K, Decker K (1977) Energy conservation in chemotrophic anaerobic bacteria. Bacteriol Rev 41: 100-180

Tsai HH, Sunderland D, Gibson GR, Hart CA, Rhodes JM (1992) A novel mucin sulphatase from human faeces: its identification, purification and characterisation. Clin Sci 82: 447-454

Turton LJ, Drucker Db, Ganguli LA (1983) Effect of glucose concentration in the growth medium upon neutral and acidic fermentation end-products of *Clostridium bifermentans, Clostridium sporogenes* and *Peptostreptococcus anaerobius*. Med Microbiol 16: 61-67

Van Soest PJ (1975) Physico-chemical aspects of fibre digestion. In: McDonald IW, Warner ACI (eds) Digestion and metabolism in the ruminant, University of New England Publishing Unit, Armidale pp 351-365

Venter CS, Vorster HH, Cummings JH (1990) Effects of dietary propionate on carbohydrate and lipid metabolism in man. Am J Clin Nutr 85: 549-553

Vercellotti JR, Salyers AA, Bullard WS, Wilkins TD (1977) Breakdown of mucin and plant polysaccharides in the human colon. Can J Biochem 55: 1190-1196

Visek WJ (1972) Effects of urea hydrolysis on cell life-span and metabolism. Fed Proc 31: 1760-1765

Visek WJ (1978) Diet and cell growth modulation by ammonia. Am J Clin Nutr 31: S216-S220

Wahren A, Holme T (1973) Amino acid and peptide requirement of *Fusiformis necrophorus*. J Bacteriol 116: 279-284

Wahren A, Bernholm K, Holme T (1971) Formation of proteolytic activity in continuous culture of *Sphaerophorus necrophorus*. Acta Pathol Microbiol Scan 79: 391-398

Walker R (1975) Naturally occurring nitrate/nitrite in foods. J Sci Food Agr 26: 1735-1742

Wallace RJ, Cheng K-J, Dinsdale D, Orskov ER (1979) An independent microbial flora of the epithelium and its role in the microbiology of the rumen. Nature 279: 424-426

Ward LA, Johnson KA, Robinson IM, Yokohama MT (1987) Isolation from swine feces of a bacterium which decarboxylates p-hydroxyphenylacetic acid to 4-methylphenol (p-cresol). Appl Environ Microbiol 53: 189-192

Weber FL, Banwell JG, Fresard KM, Cummings JH (1987) Nitrogen in fecal bacteria, fiber and soluble fractions of patients with cirrhosis: effects of lactulose and lactulose plus neomycin. J Lab Clin Med 110: 259-263

White Tabor C, Tabor H (1985) Polyamines in microorganisms. Microbiol Rev 49: 81-99

Widdel F, Pfennig N (1977) A new anaerobic sporing acetate-oxidising sulfate-reducing bacterium, *Desulfotomaculum acetoxidans* (emend). Arch Microbiol 112: 119-122

Widdel F, Pfennig N (1981a) Sporulation and further nutritional characteristics of *Desulfotomaculum acetoxidans*. Arch Microbiol 129: 401-402

Widdel F, Pfennig N (1981b) Studies on dissimilatory sulfate-reducing bacteria that decompose fatty acids. I Isolation of new sulfate-reducing bacteria enriched with acetate from saline environments. Description of *Desulfobacter potsgatei* gen nov, sp nov Arch Microbiol 129: 395-400

Widdel F, Pfennig N (1982) Studies on dissimilatory sulfate-reducing bacteria that decompose fatty acids. II Incomplete oxidation of propionate by *Desulfobulbus propionicus* gen nov, sp nov Arch Microbiol 131: 360-365

Wiggins HS (1984) Nutritional value of sugars and related compounds undigested in the gut. Proc Nutr Soc 43: 69-75

Wiggins HS, Cummings JH (1976) Evidence for the mixing of residue in the human gut. Gut 17: 1007-1011

Wilkins TD (1981) Microbiological considerations in interpretation of data obtained from experimental animals. Banbury report 7: 3-9

Wilson DR, Ing TS, Metcalfe-Gibson A, Wrong OM (1968) *In vivo* dialysis of faeces as a method of stool analysis. III The effect of intestinal antibiotics. Clin Sci 24: 211-221

Winfrey MR, Zeikus JG (1977) Effect of sulfate on carbon and electron flow during microbial methanogenesis in freshwater sediments. Appl Environ Microbiol 33: 275-281

Wolin MJ (1974) Metabolic interactions among intestinal microorganisms. Am J Clin Nutr 27: 1370-1378

Wolin MJ (1976) Interactions between $H_2$-producing and methane-producing species. In: Schlegel HG, Gottschalle G, Pfennig N (eds) Microbial formation and utilisation of gases. Goltze Press, Gottingen pp 141-150

Wolin MJ (1979) The rumen fermentation: a model for microbial interactions in anaerobic ecosystems. Adv Microbial Ecol 3: 49-77

Wolin MJ, Miller TL (1983) Carbohydrate fermentation. In: Hentges DJ (ed) Human intestinal microflora in health and disease. Academic Press, London pp 147-165

Wright DE (1967) Metabolism of peptides by rumen microorganisms. Appl Microbiol 15: 547-550

Wrong OM, Edmonds CJ, Chadwick VS (1981) The large intestine. Its role in mammalian nutrition and homeostasis. MTP Press, Lancaster

Wrong OM, Vince AJ, Waterlow JC (1985) The contribution of endogenous urea to faecal ammonia in man, determined by $^{15}N$-labelling of plasma urea. Clin Sci 68: 193-199

Yokohama MT, Carlson JR (1979) Microbial metabolites of tryptophan in the intestinal tract with special reference to skatole. Am J Clin Nutr 32: 173-178

Yokohama MT, Carlson JR (1981) Production of skatole and p-cresol by a rumen *Lactobacillus* sp Appl Environ Microbiol 41: 71-76

Yokohama MT, Tabori C, Miller ER, Hogberg MG (1982) The effects of antibiotics in the weanling pig diet on growth and excretions of volatile phenolic and aromatic bacterial metabolites. Am J Clin Nutr 35: 1417-1424

Young GP (1991) Butyrate and the molecular biology of the large bowel. In: Cummings JH, Rombeau JL, Sakata T (eds) Short chain fatty acids: metabolism and clinical importance. Ross Laboratories Press, Columbus pp 39-44

Chapter 3

# Intestinal Bacteria and Disease

G. R. Gibson and G. T. Macfarlane

## Introduction

At the beginning of this century, Metchnikoff (1907) suggested that bacteria growing in the human large intestine affected health and longevity of the host.

The majority of diseases that occur in the human large intestine are of unknown aetiology, but bacteria have been implicated either as causative agents or maintenance factors involved in many colonic disorders. A number of species are able to upset the normal gut homeostasis and cause an acute inflammatory response. The principal organisms involved are enterotoxigenic strains of *Escherichia coli*, as well as species belonging to the genera *Salmonella, Shigella, Campylobacter, Yersinia* and *Aeromonas* (Cohen and Giannella 1991). The clinical effects of these bacteria are usually short term with the patient experiencing fever, abdominal pain and diarrhoea. Acute inflammation in the colon is usually associated with ingestion of inadequately prepared or stored foods and a faecal/oral route of transmission.

The role of bacteria in other, more chronic, forms of gut disease is less clear however. Antibiotic associated colitis, inflammatory bowel disease (IBD) and colon cancer are all been thought to have an aetiology connected with the activites of the gut flora.

## Antibiotic Associated Colitis

Pseudomembranous colitis (PMC) is a severe form of colitis that occurs almost exclusively with antibiotic exposure. Initially the disease was associated with side effects of the antibiotic clindamycin (Tedesco 1976) and with *Staphylococcus aureus* infections (Cohen and Giannella 1991). However, subsequent studies revealed that *Clostridium difficile* could be isolated from almost all patients with PMC (Bartlett *et al.* 1978) and this organism is now implicated as the principal causative agent. Almost all broad spectrum antibiotics may be involved in suppression of the normal gut flora leading to PMC, but the most frequently implicated are ampicillin, clindamycin and the cephalosporins (Cohen and Gianella 1991). The disease is usually thought to arise during antibiotic therapy, with subsequent proliferation of indigenous *C. difficile* occurring. Alternatively, the patient may be exposed to environmental sources of the bacterium. *C. difficile* is detectable in the faeces of 3% of adults, but in infants the carriage may be as high as 55% (Donta and Myers 1982). Interestingly, whilst the bacterium is always associated with PMC in adults, infants can remain largely asymptomatic. The reasons for this are not as yet apparent, but have been hypothesised to be related to differences in strain virulence, a lack of toxin co-factors/receptors in children or some inhibitory mechanism towards toxin formation (Bartlett 1983; Cooperstock and Zedd 1983).

Other gastrointestinal diseases which are thought to be related to the activities of clostridia include *C. perfringens* enterotoxin-associated diarrhoea and neutropenic enterocolitis (Borriello 1985). The former disease is also associated with a range of broad spectrum antibiotics, which allow enterotoxin production by *C. perfringens*. Neutropenic entrocolitis is localised in the caecum and is characterised by wall thickening and oedema (King *et al.* 1984). It is a rapidly fatal disease in man and a number of clostridia such as *C. perfringens*, *C. sphenoides*, *C. paraperfringens* and *C. sordellii* have been isolated from patients (Felitti 1973; Gruter 1985; Newbold *et al.* 1987). However, it is now agreed that *C. septicum* is the most common causative agent (Alpern and Dowell 1969; Borriello 1985).

## Inflammatory Bowel Disease (IBD)

IBD can conveniently be divided into two major entities, namely ulcerative colitis (UC) and Crohn's disease (CD). Both conditions involve an inflammatory reaction and share many clinical features, which frequently make individual diagnosis difficult (Price, 1977). Major differences are that CD affects primarily the small gut and all regions of the large bowel, whereas UC is usually restricted to the distal colon (Whitehead 1989). Also, UC involves rectal bleeding and stool friability, whereas CD often involves the development of granulomas (Gilmour 1989). The aetiology of IBD is unknown, although various hypotheses exist.

## Ulcerative Colitis

The inflammatory response of UC is primarily located in the colonic mucosa and sub mucosa. The distal colon is always affected with the condition expressing itself in acute attacks followed by periods of symptom free remission. Bacterial involvement has been proposed in both the initiation and maintenance stages of UC (Hill 1986). *Streptococcus mobilis*, fusobacteria, and shigellas have all received attention as specific causative agents (Onderdonk 1983), largely because the organisms are either able to penetrate the gut mucosal epithelium or cause similar disease symtoms in animals. It also seems that strains of *E. coli* isolated from the colitic bowel have increased adhesive properties (Chadwick 1991). Bradley *et al.* (1987) have shown that a higher than normal proportion of facultative anaerobes and clostridia are present in patients with IBD. However, Onderdonk and Bartlett (1979) reported that antimicrobial agents active against obligate anaerobes prevented ulceration in guinea pigs and Monteiro *et al.* (1971) reported increased antibody production against anaerobes. More specifically, Roediger (1992) suggested that bacterial metabolism of sulphur compounds produces mercapto fatty acids which are able to initiate colitis lesions. In guinea pigs, feeding of the sulphated polysaccharide carrageenan leads to a development of colitis, a condition which does not occur in germ free animals (Marcus and Watt 1969; Watt and Marcus 1971). Our studies have shown an increased incidence of sulphate-reducing bacteria (SRB) in UC patients compared to healthy controls (Florin *et al.* 1990). Furthermore, production of $H_2S$ by SRB isolated from the colitic bowel is elevated and the bacteria *in vitro* are able to adapt to some clinical symptoms of the disease such as low apparent substrate availability and high wash out rates (Gibson *et al.* 1991). In general, however, evidence for a specific transmissible agent in UC is weak, since antibody production is usually low and the bacteria that have been implicated by various workers are not always found in all patients with the disease. Notably, Koch's postulates have not so far been fufilled for UC (Chadwick 1991).

More direct and convincing evidence exists for a bacterial role in disease maintenance. Onderdonk (1983) and Hill (1986) have reviewed the factors involved. Firstly, the use of drugs, particularly antibiotics to treat UC after its initiation is often effective. The most commonly used agent is sulphasalazine, which contains the antibiotic sulphapyridine (Janowitz and Bilotta 1991) and is more useful in later rather than during acute stages of the disease. Other studies have shown that metronidazole is as effective as sulphasalazine (Onderdonk *et al.*1978). It has also been reported that a recurrence of UC often occurs after acute intestinal infection. *Campylobacter jejuni, Clostridium difficile* and *Streptococcus mobilis* have all been suggested as the bacteria involved (Onderdonk 1983). However, it is probable that any non specific disturbance of the gut microflora will lead to recurrence. Reports of a role for autoimmunity and delayed type hypersensitivity, with the large intestine as the immune organ, have appeared (Monteiro *et al.* 1971). This theory would support the presence of bacterial antigens on a more long term basis. A number of bacterial products may act as inflammatory agents, such as peptidoglycan polysaccharides (Sartor *et al.* 1985), oligopeptides (Chadwick *et al.* 1988) and muramyl peptides (Bahr and Chedid 1986). More evidence for a role for bacteria in UC maintenance comes from animal experiments, such as the carrageenan model described earlier. However, certain problems with the interpretation of results from animal models exist (Chadwick 1991). For example, variations in the pathology of lesion production occur, and because there are physiological and ecological differences in the composition of the different animal gut microfloras, the immune response is bound to vary. In summary therefore, a number of studies have suggested that there is bacterial involvement in UC, yet the evidence implicating specific individual bacterial species or groups of organisms remains inconclusive at present.

## Crohn's Disease

*Eubacterium, Peptostreptococcus, Pseudomonas, Bacteroides vulgatus, C. difficile* as well as cell wall deficient L-forms of bacteria are all examples of microorganisms which have been associated with the onset of CD (Chadwick 1991). However, because the disorder involves a granulomatous reaction, it is more likely that a persistent stimulus, which macrophages have difficulty in eliminating, is involved. Mycobacteria have received considerable attention in this respect and *M. paratuberculosis* has been isolated from a number of CD patients in different centres (Chiodini *et al.* 1984; Gitnick *et al.* 1985; Graham *et al.* 1987). It is thought by some workers that disease onset is a host response to mycobacterial antigens. However, antimycobacterial chemotherapy seems to be of limited value in the treatment of the disease (Schaffer *et al.* 1984) and the organism is not always detected in CD tissue. As with UC, therefore, convincing evidence regarding the aetiology of CD is not readily apparent.

## Cancer and Colonic Bacteria

The large intestine is the second most common site for carcinoma production in man (Morotomi *et al.* 1990) with faeces from individuals living in the West frequently containing mutagenic substances as evidenced by the Ames test (Bruce and Dion 1980; Dion and Bruce 1983). Hitherto there has been no general agreement regarding the aetiology of bowel cancer, although a number of factors such as diet, environment and genetics have all been implicated (Ahnen 1991). It has been speculated that tumours occur 100 times more often in the hindgut compared to the small gut (Van Tassell *et al.* 1990) indicating that the human colonic flora may contribute towards carcinogenesis. Rowland (1988) has suggested that a major mechanism by which the gut flora may be involved is by the production of carcinogens from non-toxic precursor molecules. A number of groups are currently investigating the role of colonic bacteria in carcinogen production. The majority of studies involve the incubation of pure or mixed cultures of bacteria with precursor compounds of interest and subsequently monitoring potential carcinogen formation. These may then be followed up by experiments comparing metabolism in conventional and germ-free animals. In this way, a number of bacterial mechanisms which contribute to tumour production in the colon, or elsewhere in the body have been postulated (see Table 3.1). However as yet, none of these mechanisms have been proven to occur significantly *in vivo* and it is more than likely that a number of interacting factors are necessary to initiate, promote and develop tumour formation. Here, we will summarise some of the bacterial mechanisms which may contribute towards carcinogen production.

Fecapentaenes are conjugated ether lipids which exert potent mutagenic activity (Van Tassell *et al.* 1990). It has been suggested that these compounds are the most prevalent genotoxins in the colon and are excreted by 75% of persons consuming a typical Western diet (Schiffman *et al.* 1988; Van Tassel *et al.* 1989). Species belonging to the genus *Bacteroides* are able to produce significant amounts of fecapentaene *in vitro*, the production rate being greatly enhanced by the addition of bile to the culture system (Van Tassel *et al.* 1982a,b). Precursors for this reaction are polyunsaturated ether phospholipids (Van Tassell *et al.* 1990).

Dietary intakes of drinking water and other foodstuffs may augment the supply of nitrate available in the colon from endogenous origins (Witter *et al.* 1979; Radcliffe *et al.* 1985; Iyengar *et al.* 1987). After reduction to nitrite by intestinal microorganisms, N-nitrosation of secondary amines may occur to produce potentially harmful nitrosamines (Archer *et al.* 1981). These compounds have adverse effects towards colonic epithelial cells and are potently carcinogenic. N-Nitrosation occurs optimally at acid pH and nitrosamine formation may be of some importance in the stomach. More interestingly however, some intestinal bacteria are capable of forming N-nitrosamines at a neutral pH (Suzuki and Mitsuoka 1984; Calmels *et al.* 1985). A number of reports have indicated that nitrosamines are able to produce tumours at

**Table 3.1.** Potentially carcinogenic compounds produced by bacteria growing in the human large intestine

| Carcinogen | Mechanism of production | Reference |
|---|---|---|
| Fecapentaenes | Plasmalogens (polyunsaturated ether lipids) are converted to mutagenic fecapentaenes by some *Bacteroides* spp. | Van Tassel *et al.* 1989; 1990<br>Kingston *et al.* 1990 |
| Nitrosamines | Combination of nitrite with a secondary or tertiary amine | Bruce *et al.* 1979<br>Bruce and Dion 1980<br>Suzuki and Mitsuoka 1984 |
| Bile acids | Bacterial metabolism into co-carcinogens e.g. steroids | Hill *et al.* 1971<br>Thompson 1982 |
| Heterocyclic amines | Mutagenic after activation by S9 microbial enzymes | Kasai *et al.* 1980a,b<br>Bashir *et al.* 1987<br>Van Tassell *et al.* 1990 |
| Phenolic compounds | Co-carcinogens produced from tyrosine metabolism | Bone *et al.* 1976 |
| Various aglycones | Microbial ß-glycosidase activity towards glycoside conjugates e.g. hydrolysis of cycasin by ß-glucosidase | Spatz *et al.* 1967<br>Tamura *et al.* 1980<br>Branscomb *et al.* 1988 |
| Azo dyes | Possibly carcinogenic after reduction by intestinal anaerobes | Chung *et al.* 1978<br>Rafii *et al.* 1990 |
| Glucuronide compounds | Microbial ß-glucuronidase activity | Clark *et al.* 1969 |
| Diacylglycerol | Metabolism of phosphatidyl choline (dietary lipid) in conjunction with bile acids | Morotomi *et al.* 1990 |
| Nitrated polycyclic aromatic hydrocarbons (nitro-PAH's) | Bacterial nitro reduction causes carcinogenesis in laboratory animals | Fu *et al.* 1988<br>Manning *et al.* 1988<br>Rafii *et al.* 1991 |
| Ammonia | Bacterial deamination of amino acids and nitrate reduction | Macfarlane *et al.* 1986<br>Allison and Macfarlane 1988 |
| Amines | Mainly decarboxylation of aromatic amino acids | Gale 1946<br>Goldschmidt *et al.* 1971 |
| Indolic compounds | Co-carcinogens produced from tryptophan metabolism | Williams 1972<br>Renwick 1986 |

various body sites in rats but the liver seems to be primarily targetted (see Goldin 1986).

Hill and his co-workers have suggested that faecal bile acid concentration correlates with the risk of large bowel cancer in various populations (Aries *et al.* 1969; Hill *et al.* 1971, 1975; Drasar and Hill 1974). It is thought that colonic bacteria convert bile acids into steroids, which have carcinogenic properties (Aries *et al.* 1971). Hill *et al.* (1971) studied faecal bacterial populations in persons in high (England, Scotland, USA) and low (India, Uganda, Japan) risk areas for colonic cancer. They found that faeces from the low risk populations contained lower populations of bacteroides and bifidobacteria, with increased counts of faecal streptococci and a low anaerobe:aerobe ratio. Moreover, the ability of anaerobic bacteria to dehydroxylate cholic acid was greater in the high risk areas. However, Vargo *et al.* (1980) postulated that a predominance of aerobic bacteria, and a decrease in anaerobic cocci, *Eubacterium* spp. and *Fusobacterium* spp. occurred in a group of colon cancer patients when compared to persons with non-malignant disease. Care must be taken when interpreting results from such studies however, as apparent changes in the microflora are not necessarily related to disease initiation, but may result from effects of the disease itself and consequently, unreliable information may be obtained using faecal samples where the normal gut homeostasis has been upset. The situation is further confused by considering the results of Moore and Holdeman (1975) and Finegold *et al.* (1975) who were unable to demonstrate significant differences in the composition of the faecal flora in persons from high and low risk areas.

A heterocyclic amine, 2-amino-3-methyl-3H-imidazo [4,5-f] quinoline (IQ) has been isolated by Kasai *et al.* (1980a,b) from broiled fish. This compound is also found in beef extract (Felton *et al.* 1984). After activation by the liver, IQ is highly mutagenic (Barnes *et al.* 1983). It is thought that IQ and other heterocyclic amines may be involved in the formation of colorectal cancers (Van Tassel *et al.* 1990). Carmen *et al.* (1987) showed that IQ could also be metabolised to a harmful 7-hydroxy form by mixed faecal bacteria and pure cultures of eubacteria and clostridia. Importantly, 7-hydroxy IQ, if formed by the gut microflora, would not need to undergo enterohepatic circulation before exerting its effect on DNA synthesis (Van Tassell *et al.* 1990).

A number of conversion reactions whereby bacterial enzymes are able to form carcinogens from dietary components have received attention (Wilkins and Van Tassell 1983). For example, bacterial ß-glucosidase hydrolyses glucoside conjugates such as cycasin (methylazoxymethanol ß-D-glucoside) to yield the aglycone methylazoxymethanol which is carcinogenic in rats (Spatz *et al.* 1967), whilst strains of *Streptococcus* have been shown to hydrolyse rutin to produce the mutagen quercetin (MacDonald *et al.* 1984). Similarly, microbial ß-glucuronidase is able to hydrolyse glucuronides allowing steroid reabsorption (Clark *et al.* 1969). Interestingly, analysis of gut contents from sudden death victims (Macfarlane *et al.* 1991) and faeces, indicate only low ß-glucuronidase activities throughout the colon, whereas, ß-glucosidase levels are considerably greater.

Recently, Morotomi *et al.* (1990) proposed a new hypothesis to explain the role of dietary lipids and the intestinal microflora in the aetiology of colon cancer. The incubation of faecal bacteria with phosphatidylcholine in association with deoxycholic acid, produced diacylglycerol which activates protein kinase C, a key enzyme involved in the control of growth. Significant activities of this enzyme were found in cultures of *Clostridium perfringens*.

Other enzymes produced by bacteria that may be involved in carcinogen production include nitroreductases and azoreductases. The former group are able to metabolise nitroaromatic compounds, which are common in the environment, to form aryl hydroxyamines which can undergo esterification and interact with DNA (Richardson *et al.* 1988). Azo dyes are frequently added as colouring additives used in the food industry and are not well absorbed in the small gut. The bacterial flora of the hindgut is able to reductively hydrolyse the azo bond to produce in some cases, toxic aromatic amines (Chung *et al.* 1975). The main bacteria implicated are species belonging to the genera *Eubacterium* and *Clostridium* (Rafii *et al.* 1990).

From the foregoing, it would appear that colonic bacteria exert a solely detrimental effect towards host welfare in terms of tumour production. However, the situation is much more complex than this and many species are undoubtedly involved in detoxification processes. For example, many of the purported protective effects of dietary fibre can be attributed to the activities of the gut microflora (Cummings 1985; Lupton 1991). In particular, the production of butyrate has received attention as a regulator of gene expression and cell growth. This SCFA is able to prolong cell growth rate, stabilise chromatin structure during cell division and reduce DNA synthesis *in vitro* (Hagopian *et al.* 1977; Prasad 1980; Kim *et al.* 1982).

As we have seen, the colonic microflora may be directly or indirectly involved in the initiation of gut disease. However, the specific organisms involved, their mechanisms of action and significance *in vivo* have yet to be positively identified. In many ways, it is important that this problem be addressed before effective preventative action can be identifed. There is some evidence that optimisation of the composition and activities of the gut microflora may result in a significant improvement in host health. For example, an increase in the numbers of certain purportedly beneficial groups of bacteria (e.g. lactobacilli, bifidobacteria) may be achievable by dietary manipulation. Alternatively, microflora replacement therapy whereby faecal enemas are administered is also possible. In this context, preliminary results reported by Bennet and Brinkmann (1989) seem encouraging for the alleviation of some of the symptoms of ulcerative colitis.

## References

Ahnen DJ (1991) Etiology of large bowel cancer. In: Phillips SF, Pemberton JH, Shorter RG (eds) The large intestine: physiology, pathophysiology and disease. Raven Press Ltd, New York pp 501-520

Allison C, Macfarlane GT (1988) Effect of nitrate on methane production by slurries of human faecal bacterial. J Gen Microbiol 134: 1397-1405

Alpern RJ, Dowell VR (1969) *Clostridium septicum* infections and malignancy. J Am Med Assoc 209: 385-389

Archer MC, Saul RL, Lee LJ, Bruce WR (1981) Analysis of nitrate, nitrite and nitrosamines in human feces. In: Bruce WR, Correa P, Lipkin M, Tannenbaum SRR (eds) Banbury report no 7. Cold Spring Harbor Laboratory, New York pp 321-327

Aries VC, Crowther JS, Drasar BS, Hill MJ, Ellis FR (1971) The effect of a strict vegetarian diet on the faecal flora and faecal steroid concentration. J Pathol 103: 54-56

Aries VC, Crowther JS, Drasar BS, Hill MJ, Williams REO (1969) Bacteria and the etiology of cancer of the large bowel. Gut 10: 334-335

Bahr GM, Chedid L (1986) Immunological activites of muramyl Barker HA (1981) Amino acid degradation by anaerobic bacteria. Ann Rev Biochem 50: 23-40

Barnes WS, Mailler J, Weisburger JH (1983) *In vitro* binding of the food mutagen 2-amino-3-methylimidazo [4,5-f] quinoline (IQ) and analogs. Carcinogenesis 6: 441-444

Bartlett JG (1983) Pseudomembranous colitis. In: Hentges DJ (ed) Human intestinal microflora in health and disease. Academic Press, London pp 448-479

Bartlett JG, Chang TW, Gurwith M, Gorbach SL, Onderdonk AB (1978) Antibiotic-associated pseudomembranous colitis due to toxin producing clostridia. N Engl J Med 298: 531-534

Bashir M, Kingston DGI, Carman RL, Van Tassell RL, Wilkins TD (1987) Anaerobic metabolism of 2-amino-3-methyl-3H-imidazo [4,5-f] quinoline (IQ) by human fecal flora. Mutat Res 190: 187-190

Bennet JD, Brinkman M (1989) Treatment of ulcerative colitis by implantations of normal colonic flora. Lancet 164

Bone E, Tamm A, Hill MJ (1976) The production of urinary phenols by gut bacteria and their possible role in the causation of large bowel cancer. Am J Clin Nutr 29: 1448-1454

Borriello SP (1985) Newly described clostridial diseases of the gastrointestinal tract: *Clostridium perfringens* enterotoxin-associated diarrhea and neutropenic enterocolitis due to *Clostridium septicum*. In: Borriello SP (ed) Clostridia in gastrointestinal disease. CRC Press, Boca Raton pp 223-229

Bradley HK, Wyatt GM, Bayliss CE, Hunter JO (1987) Instability in the faecal flora of a patient suffering from food-related irritable bowel syndrome. Med Microbiol 23: 29-32

Branscomb CJ, Holder CL, Kofmacher WA, Cerniglia CE, Rushing LG (1988) GC/MS characterisation of urinary metabolites of deoxyalanine succinate: identification of aglycones formed from intestinal microflora metabolism of the polar glucuronide metabolites. J High Res Chromat and Chromat Comm 11: 517-520

Bruce WR, Dion PW (1980) Studies relating to a fecal mutagen. Am J Clin Nutr 33: 2511-2512

Bruce WR, Varghese AJ, Wang S, Dion P (1979) The endogenous production of nitroso compounds in the colon and cancer at that site. In: Miller EC (ed) Naturally occurring carcinogens, mutagens and modulators of carcinogenesis. Japan Sci Press, Tokyo pp 221-228

Calmels S, Ohshima H, Vincent P, Gounat AM, Bartsch H (1985) Screening of microorganisms for nitrosation catalysis at pH7 and kinetic studies on nitrosamine formation from secondary amines by *Escherichia coli* strains. Carcinogenesis 6: 911-916

Carmen RJ, Van Tassell RL, Kingston DGI, Bashir M, Wilkins TD (1987) Conversion of IQ, a dietary pyrolysis carcinogen, to a direct acting mutagen by normal intestinal bacteria of humans. Mutat Res 206: 335-342

Chadwick VS (1991) Etiology of chronic ulcerative colitis and Crohn's disease. In: Phillips SF, Pemberton JH, Shorter RG (eds) The large intestine: physiology, pathophysiology and disease. Raven Press Ltd, New York pp 445-463

Chadwick VS, Mellor DM, Myers DB (1988) Production of peptides inducing chemotaxis and lysosomal enzyme release in human neutrophils by intestinal bacteria *in vitro* and *in vivo*. Scan J Gastroenterol 23: 121-128

Chiodini RJ, Van Kruningen HJ, Thayer WR, Merkal RS, Coutu JA (1984) Possible role of mycobacteria in inflammatory bowel disease. 1 An unclassified *Mycobacterium* species isolated from patients with Crohn's disease. Dig Dis Sci 29: 1073-1079

Chung KT, Folk GE, Egan M (1978) Reduction of azo dyes by intestinal anerobes. Appl Environ Microbiol 35: 558-562

Chung KT, Folk GE, Slein MW (1975) Tryptophanase of fecal flora as a possible factor in the etiology of colon cancer. J Nat Can Ins 54: 1073-1078

Clark AG, Fischer LG, Millburn P, Smith PL, Williams RT (1969) The role of the gut flora in the enterohepatic circulation of stilboestrol in the rat. Biochem J 112: 17P

Cohen MB, Giannella RA (1991) Bacterial infections: pathophysiology, clinical features and treatment. In: Phillips SF, Pemberton JH, Shorter RG (eds) The large intestine: physiology, pathophysiology and disease. Raven Press Ltd, New York pp 395-428

Cooperstock MS, Zedd AJ (1983) Intestinal flora of infants. In: Hentges DJ (ed) Human intestinal microflora in health and disease. Academic Press, London pp 79-99

Cummings JH (1985) Cancer of the large bowel. In: Trowell H, Burkitt D, Heaton K (eds) Dietary fibre, fibre depleted foods and disease. Academic Press, London pp 161-189

Dion P, Bruce WR (1983) Mutagenicity of different fractions of extracts of human feces. Mutat Res 119: 151-160

Donta ST, Myers MG (1982) *Clostridium difficile* toxin in asymptomatic neonates. J Pediatr 10: 431-434

Drasar BS, Hill MJ (1974) Human intestinal flora. Academic Press, London

Fellitti VJ (1973) Primary invasion by *Clostridium sphenoides* in a patient with periodic neutropenia. Calif Med 133: 76-78

Felton JS, Knize MG, Wood C, Wuebbles BJ, Heally KS, Stuermer DH, Bjeldanes LF, Kimble BJ, Hatch FT (1984) Isolation and characterisation of new mutagens from fried ground beef. Carcinogenesis 5: 95-102

Finegold SM, Flora DJ, Attlebury HR, Sutter LV (1975) Fecal bacteriology of colonic polyp patients and control patients. Cancer Res 35: 3407-3417

Florin THJ, Gibson GR, Neale G, Cummings JH (1990) A role for sulfate-reducing bacteria in ulcerative colitis? Gastroenterology 98: A170

Fu PP, Cerniglia CE, Richardson KE, Heflich RH (1988) Nitroreduction of 6-nitrobenzo[a]pyrene: a potential activation pathway in humans. Mutat Res 209: 123-129

Gale EF (1946) The bacterial amino acid decarboxylases. Adv Enzymol 6: 1-32

Gibson GR, Cummings JH, Macfarlane GT (1991) Growth and activities of sulphate-reducing bacteria in gut contents of healthy subjects and patients with ulcerative colitis. FEMS Microbiol Ecol 86: 103-112

Gilmour AM (1989) Crohn's disease. In: Whitehead R (ed) Gastrointestinal and oesophageal pathology. Churchill Livingstone, Edinburgh, p 540

Gitnick G, Collins J, Beaman B, Brooks D, Arthur M (1985) Mycobacteria in Crohn's disease. Gastroenterology 88: A1592

Goldin B (1986) In situ bacterial metabolism and colon mutagens. Ann Rev Microbiol 40: 367-393

Goldschmidt MC, Lockhart BM (1971) Rapid methods for determining decarboxylase activity: arginine decarboxylase. Appl Microbiol 22: 350-357

Graham DY, Markesich DC, Yoshimura HH (1987) Mycobacteria and inflammatory bowel disease. Results of culture. Gastroenterology 92: 436-442

Gruter H (1985) Gas gangrene following antibiotic-associated enterocolitis in hereditary neutropenia. Arch Anat Cytol Pathol 33: 23-25

Hagopian HK, Riggs MG, Swartz LA, Ingram VM (1977) Effect of n-butyrate on DNA synthesis in chick fibroblasts and hela cells. Cell 12: 855-860

Hill MJ (1986) The possible role of bacteria in inflammatory bowel disease. Curr Con Gastroenterol 3: 10-14

Hill MJ, Drasar BS, Aries V, Crowther JS, Williams REO (1971) Bacteria and aetiology of cancer of the large bowel. Lancet 95-100

Hill MJ, Drasar BS, Williams REO, Meade TW, Cox AG, Simpson JEP, Morson BC (1975) Faecal bile acids and clostridia in patients with cancer of the large bowel. Lancet 535-538

Iyengar R, Stuele DJ, Marletta MA (1987) Macrophage synthesis of nitrite, nitrate and N-nitrosamines: precursors and role of the respiratory burst. Proc Natl Acad Sci USA 84: 6397-6373

Janowitz HD, Bilotta JJ (1991) Critical evaluation of the medical therapy of inflammatory bowel disease. In: Phillips SF, Pemberton JH, Shorter RG (eds) The large intestine: physiology, pathophysiology and disease. Raven Press Ltd, New York pp 475-500

Kasai HZ, Nishimura S, Wakabayashi K, Nagao M, Sugimura T (1980a) Chemical synthesis of 2-amino-3-methylimidazole [4,5-f] quinoline (IQ), a potent mutagen isolated from broiled fish. Proc Jpn Acad 56: 382-384

Kasai HZ, Yamauimi Z, Wakabayashi K, Nagao M, Sugimura T (1980b) Potent novel mutagens produced by broiling fish under normal conditions. Proc Jpn Acad 56: 278-283

Kim YS, Tsao D, Morita A, Bella A (1982) Effect of sodium butyrate and three human colorectal adenocarcinoma cell lines in culture. Falk Symp 31: 317-323

King A, Rampling A, Wright DGD, Warren RE (1984) Neutropenic enterocolitis due to *Clostridium septicum* infection. J Clin Pathol 37: 335-343

Kingston DGI, Van Tassel RL, Wilkins TD (1990) The fecapentaenes, potent mutagens from human feces. Chem Res Toxicol 3: 391-400

Lupton JR (1991) Dietary fiber and short chain fatty acids - implications for colon cancer: animal models. In: Cummings JH, Rombeau JL, Sakata T (eds) Short chain fatty acids: metabolism and clinical importance. Ross Laboratories Press, Columbus pp 86-90

MacDonald IA, Bussard RG, Hutchinson DM, Holdeman LV (1984) Rutin-induced ß-glucosidase activity in *Streptococcus faecium* VGH-1 and *Streptococcus* sp. strain FRP-17 isolated from human feces. Formation of the mutagen quercetin from rutin. Appl Environ Microbiol 47: 350-355

Macfarlane GT, Cummings JH, Allison C (1986) Protein degradation by human intestinal bacteria. J Gen Microbiol 132: 1647-1656

Macfarlane GT, Gibson GR, Cummings JH (1991) Extracellular and cell-associated glycosidase activities in different regions of the human large intestine. Lett Appl Microbiol 12, 3-7

Manning BW, Campbell WL, Franklin W, Declos KB, Cerniglia CE (1988) Metabolism of 6-nitrochrysene by intestinal microflora. Appl Environ Microbiol 54: 197-203

Marcus R, Watt J (1969) Seaweeds and ulcerative colitis in laboratory animals. Lancet 489-490

Metchnikoff E (1907) The prolongation of life. Heinemann, London

Monteiro E, Fossey J, Shiner M, Drasar BS, Allison AC (1971) Antibacterial antibodies in rectal and colonic mucosa in ulcerative colitis. Lancet 249-251

Moore WEC, Holdeman LV (1975) Discussion of current bacteriological investigations of the relationships between intestinal flora, diet, and colon cancer. Cancer Res 35: 3418-3420

Morotomi M, Guillem JG, LoGerfo P, Weinstein IB (1990) Production of diacylglycerol, an activator of protein kinase C, by human intestinal microflora.Cancer Res 50: 3595-3599

Newbold KM, Lord MG, Baglin TP (1987) Role of clostridial organisms in neutropenic enterocolitis. J Clin Pathol 40: 471

Onderdonk AB (1983) Role of the intestinal microflora in ulcerative colitis. In: Hentges DJ (ed) Human intestinal microflora in health and disease. Academic Press, London pp 481-493

Onderdonk AB, Bartlett MD (1979) Bacteriological studies of experimental ulcerative colitis. Am J Clin Nutr 32: 258-265

Onderdonk AB, Hermos JA, Dzink JL, Barlett JG (1978) Protective effect of metronidazole in experimental ulcerative colitis. Am J Clin Nutr 32: 1819-1825

Prasad KN (1980) Butyric acid: a small fatty acid with diverse biological functions. Life Sci 27: 1351-1358

Price AB (1977) Difficulties in the differential diagnosis of ulcerative colitis and Crohn's disease. In: Yardley H, Morson BC (eds) The gastrointestinal tract, Williams and Wilkins, Baltimore pp 1-14

Radcliffe BC, Nance SH, Deakin FJ, Roediger WEW (1985) Nitrate and nitrite content of ileostomy effluent after a normal or high nitrate meal. Clin Invest Med 8: A94

Rafii F, Franklin W, Cerniglia CE (1990) Azoreductase activity of anaerobic bacteria isolated from human intestinal microflora. Appl Environ Microbiol 56: 2146-2151

Renwick 1986 Gut bacteria and the metabolism of aromatic amino acids. In: Hill MJ (ed) Microbial metabolism in the digestive tract. CRC Press, Boca Raton, pp 107-122

Richardson KE, Fu PP, Cerniglia CE (1988) Metabolism of 1-, 3-, and 6-nitrobenzo[a]pyrene by intestinal microflora. J Toxicol Environ Health 23: 527- 537

Roediger WEW (1992) The role of sulphur metabolism and mercapto fatty acids in the aetiology of ulcerative colitis. In: Goebell H, Ewe H, Malchow H (eds) Inflammatory bowel diseases. Progress in basic research and clinical implications, MTP Press, Lancaster - in press

Rowland IR (1988) Interactions of the gut microflora and the host in toxicology. Toxicol Pathol 16: 147-153

Sartor RB, Cromartie WJ, Powell DW, Schwab JH (1985) Granulomatous enterocolitis induced in rats by purified bacterial cell wall fragments. Gastroenterology 89: 587-595

Schaffer JL, Hughes S, Linaker BD, Baker RD, Turnberg LA (1984) Controlled trial of rifampicin and ethambutol in Crohn's disease. Gut 25: 203-205

Schiffman MH, Bitterman P, Viciana AL, Schairer C, Russell L, Van Tassell RL, Wilkins TD (1988) Fecapentaenes and their precursors throughout the bowel - an autopsy study. Mutat Res 208: 9-15

Spatz M, Smith DWE, McDaniel EG, Laquer GL (1967) Role of intestinal microorganisms in determining cycasin toxicity. Prc Soc Biol Med (New York) 124: 691-697

Spoelstra SF (1977) Simple phenol and indoles in anaerobically stored piggery wastes. J Sci Food Agr 28: 415-423

Suzuki K, Mitsuoka T (1984) N-nitrosamine formation by intestinal bacteria. In: O'Neill IK, von Borstel RC, Miller CT, Long J, Bartsch H (eds) Nitroso compounds: occurrence, biological effects and relevance to human cancer. Int Agency Res Cancer, Lyon pp 275-281

Tamura G, Gold C, Ferro-Luzzi A, Ames BN (1980) Fecalase: a model for activation of dietary glycosides to mutagens by intestinal flora. Proc Natl Acad Sci (USA) 77: 4961-4965

Tedesco FJ (1976) Clindamycin-associated colitis. Review of the clinical spectrum of 47 cases. Dig Dis 21: 26-32

Thompson MH (1982) The role of diet in relation to faecal bile acid concentration and large bowel cancer. In: Malt RA, Williamson RCN (eds) Colonic carcinogenesis. MTP Press, The Hague pp 49-56

Van Tassell RL, Kingston DGI, Wilkins TD (1990) Metabolism of dietary genotoxins by the human colonic microflora: the fecapentaenes and heterocyclic amines. Mutat Res 238: 209-221

Van Tassell RL, MacDonald DK, Wilkins TD (1982a) Production of a fecal mutagen by *Bacteroides* spp. Infect Imm 37: 975-980

Van Tassell RL, MacDonald DK, Wilkins TD (1982b) Stimulation of mutagen production in human feces by bile and bile acids. Mutat Res 103: 233-239

Van Tassell RL, Piccariello T, Kingston DGI, Wilkins TD (1989) The precursors of fecapentaenes: purification and properties of a novel plasmologen. Lipids 24: 454-459

Vargo D, Moskovitz M, Flock MH (1980) Faecal bacterial flora in cancer of the colon. Gut 21: 701-705

Watt J, Marcus R (1971) Carrageenan induced ulceration of the large intestine in the guinea pig. Gut 12: 164-171

Whitehead R (1989) Ulcerative colitis. In: Whitehead R (ed) Gastrointestinal and oesophageal physiology, Churchill Livingstone, Edinburgh pp 522-531

Wilkins TD, Van Tassell RL (1983) Production of intestinal mutagens. In: Hentges DJ (ed) Human intestinal microflora in health and disease. Academic Press, London pp 265-288

Williams RT (1972) Toxicological implications of biotransformation by intestinal microflora. Toxicol Appl Pharm 23: 769-781

Witter JP, Balish E, Gatley SJ (1979) Distribution of nitrogen-13 from labelled nitrate and nitrite in germ-free and conventional flora rats. Appl Environ Microbiol 38: 870-878

Chapter 4

# Probiotics: an Overview

R. Fuller

## Introduction

The first recorded probiotics were fermented milks produced for human consumption. However, the subsequent development of the concept has been based on results obtained in animal experiments and most of the current market in probiotics is for animal preparations (Lloyd-Evans 1989). It is, therefore, even in a contribution to a book on human probiotics, necessary to use animal data to illustrate the history of the concept and to establish the principles on which it is based and which will govern future development. It is possible to relate directly some of the animal data to humans, but other information obtained from animal experiments should be transposed with caution. Indeed, the host-specific nature of microbial gut colonisation makes it unwise to transpose results between any animal species without considering very carefully the different factors which may be operating.

## Definition

Over the course of many years practices have grown up which could be embraced by the term probiotic. Transfer of rumen contents from the mother cow to the calf and fermentation of milk, meat and fish were all used long before the word probiotic was coined.

It is, therefore, important at this point to assess the present meaning of the word. Probiotic has had several meanings and even in its present context it can mean different things to different people. It was first used to mean the opposite of antibiotic i.e. substances secreted by one micro-organism which stimulated another (Lilley and Stillwell 1965). Although this was grammatically correct it never caught on and it was subsequently used to describe tissue extracts which stimulated micro-organisms (Sperti 1971). It was not until three years later that the word probiotics was used in more or less the way in which it will be used in this book. Parker (1974) defined it as "organisms and substances which contribute to intestinal microbial balance". I felt that the use of the word "substances" was too vague and could even include antibiotics, a class of substances not intended to be embraced by the definition. I, therefore, modified it to read "a live microbial feed supplement which beneficially affects the host animal by improving its intestinal microbial balance". (Fuller 1989). This definition stresses the importance of the viability of the supplement. It includes bacteria, yeasts, moulds and bacteriophage which have all been shown to have effects on disease resistance, nutrition and growth. It includes not only preparations specifically designed to be used as feed supplements with beneficial effects but also covers fermented milks which were developed thousands of years ago quite empirically, but which had associated effects on health either directly or indirectly by prevention of spoilage.

It may become obvious during the course of this book that this definition is, or shortly will become, inadequate but for the moment I think it is a worthwhile working definition around which to structure this chapter. It is imperfect because it restricts probiotics to the intestine whereas the approach can be applied to the microbial ecology of the vagina, skin and respiratory tract. It also fails to include non-viable preparations which appear to have their effect by stimulating specific groups of bacteria associated with the animal's body.

## History and Development

Although there are earlier reports of fermented milks with implied health benefits, Metchnikoff (1907b) was the first person to put the subject on a scientific basis. He had *in vitro* evidence that *Vibrio cholerae* could be either stimulated or inhibited by other bacteria and so the outcome of a *V. cholerae* infection depended on the composition of the gut flora which it encountered after being ingested. He also believed that the flora of the large intestine was responsible for the chronic production of toxins - the so-called autointoxication - and felt that these adverse effects could be ameliorated by altering the composition of the indigenous gut microflora. This he proposed was possible by eating fermented milks. He found support for his theory in the fact that Bulgarian peasants ingested large quantities of fermented milks and also lived to great ages. He was convinced that the two observations were causally related.

This theory was developed in several publications during the early part of this century (Bibel 1988) but the publication, which gave his theories world-wide exposure and which is so often quoted to authenticate his work, is his book published in 1907(a). The original French edition was entitled "Essais optimistes". This book contained a wide-ranging discussion of the philosophy and physiology of the ageing process with two chapters devoted to the intestinal microflora and the potential benefits derived from ingesting fermented milks. Unfortunately, when the book was translated into English the title was changed to "The Prolongation of Life" with "optimistic studies" relegated to subtitle status. As a result Metchnikoff was to be forever linked in the public mind with advocating the ingestion of fermented milks to increase your life span. This seems unfair because careful reading of the book reveals that by this time (whatever he had said before) his statements on the subject were very guarded. At the end of his chapter on "Lactic acid as inhibiting intestinal putrefaction", after making a plea for more research on the effect of diet on prevention of intestinal putrefaction, he concludes:

"In the meantime those who wish to make the cycle of life as complete and as normal as is possible under present conditions must depend upon sobriety and on conforming to the rules of personal hygiene".

Perhaps his most significant contribution to the subject was his realisation that there were beneficial micro-organisms in the intestinal tract and that not all bacteria were harmful. As he puts it:

"A reader who has little knowledge of such matters may be surprised by my recommendation to absorb large quantities of microbes, as a general belief is that microbes are harmful. This belief is erroneous. There are many useful microbes, amongst which the lactic bacilli have an honourable place".

So wrote Metchnikoff in 1907(b), but it is an admonition that could well be repeated today for there is still a tendency for many people to equate micro-organisms with pathogens and asssume that no good could ever come from associating with them.

We should also note that the soured milk which Metchnikoff used is not the yoghurt which we know today consisting of milk fermented with the two starter organisms *Lactobacillus delbruechii* subsp *bulgariaicus* and *Streptococcus salivarius* subsp. *thermophilus*. Although Metchnikoff did isolate and use a lactobacillus which he referred to as the Bulgarian bacillus, its relationship to the present lactobacillus starter is not clear. Indeed, it is impossible to know with any certainty which species Metchnikoff and his contemporaries were using, but it is likely that mixtures of lactobacilli were sometimes employed unintentionally (Bibel, personal communication).

Whatever the microbiological basis of the product, the habit of consuming fermented milks continued unabated. The research interest now became centred on the best type of organism to use. It had already been suggested that intestinal isolates might be a better proposition (Rotch and Kendall 1911; Orla-Jensen 1912) and it was shown that *L. acidophilus* colonised the lower gut better than did *L. bulgaricus*. This supported the belief that an intestinal strain would colonise the gut and exert its effect over a longer period and it was later demonstrated that in order for *L. acidophilus* to be effective it had to be viable and that using such a viable preparation the effect would persist for six months after the administration ceased (Kopeloff 1923). Recent work would not necessarily support the persistence of the effect but this may vary according to the strain used and the way in which the effect is monitored.

*L. acidophilus* was used successfully to treat constipation, diarrhœa and colitis (Rettger and Chaplin 1921) although some groups were less than enthusiastic about the results (Bassler and Lutz 1922). The treatment was sometimes combined with colonic irrigation (Hughens 1925) or dietary manipulation (Hull and Rettger 1917). Experimental trials continued throughout the 1920s (see Frost and Hankinson 1931).

By the early 1930s the emphasis of research had shifted to chemotherapy and the discovery of the sulpha drugs in mid 1930s tended to dominate the thinking behind research into therapy of intestinal disease. When penicillin became a clinical possibility in 1940, this and other antibiotics became the treatment of choice.

However, in 1946 even Florey, who had been instrumental in developing penicillin as a clinical drug, was talking about "The use of organisms for therapeutic purposes". Research on the scientific basis of the health benefits of fermented milks also received a boost from antibiotics in another rather indirect way. Soon after Florey's lecture, it was demonstrated that low levels of antibiotics included in the feed of farm animals improved their growth. It became important to know how this was achieved and this required a more detailed knowledge of the gastrointestinal microflora and its effects on the host animal. The ensuing research revealed that the gut flora was a very complex mixture of micro-organisms. Not only did it contain the facultative anaerobes and the easily cultured strict anaerobes which had been described up until this time, but the development of new techniques for the isolation of anaerobes showed that there was a very important component of the gut flora which had previously not been considered.

It also became obvious that *L. acidophilus* was just one of several different species of lactobacillus that occurred in the gut together with other related lactic acid bacteria such as streptococci and bifidobacteria. Bifidobacteria had long been thought of as important intestinal micro-organisms; Tissier (1905) had described them as a dominant group in the intestine of breast-fed babies and they were subsequently considered to be responsible for the improved resistance to disease of breast-fed babies compared with formula fed babies (Bullen *et al.* 1975). Later research has shown that the viable count of bifidobacteria is the same in both types of baby (Benno and Mitsuoka 1986; Hall *et al.* 1990) but the qualitative difference described by Neut *et al.* (1980) may be important in this respect.

The lactic acid bacteria became some of the most commonly used organisms in probiotic preparations. Other micro-organisms currently being used in fermented milks or probiotic preparations include *Bacillus* sp and fungi. There is a wide range of bacteria and some yeasts and moulds; many are organisms traditionally associated with the intestinal tract.

Not included in this list are the non-pathogenic mutants of pathogens which have been used successfully to treat infections caused by *Salmonella typhimurium* in chickens (Barrow *et al.* 1987) *Clostridium difficile* in human adults (Borriello and Barclay 1985) and *Escherichia coli* in human babies (Duval-Iflah *et al.* 1982). Although they conform to the description of probiotics they are not in the same line of development as the other organisms.

## Use of Probiotics

Although fermented milks were first developed for human consumption, most of the experimental work has been done on laboratory rodents and farm animals. The preparations recently developed have been mainly for farm animals although producers often recommend them for horses and pets (cats and dogs) as well. There has also been some use of probiotics in fish farming (Robertson *et al.* 1990). The current state of the probiotic market has been extensively reviewed by Lloyd-Evans (1989).

Some attempt has been made to capitalise on the research results relating to colonisation of the gut and to design host-specific preparations with different species or strains of micro-organisms for different species of host consumer. However, the continuous administration frequently recommended by manufacturers does make the host specific approach less important.

The probiotics presently on the market come in various forms such as powders, tablets, capsules, liquid suspensions, pastes and sprays. The way in which a probiotic preparation is prescribed depends to some extent on what species of animal is to receive it. For example individual administration of pastes to cows or pigs may be possible but for a battery house full of chickens the probiotic must be included in the food or water or sprayed into the atmosphere when the chicks are in the incubator or delivery boxes.

The acceptance of probiotics for human consumption has been faster and more extensive in many countries outside the UK. Continental Europe, particularly the Scandinavian countries have a long history of fermented milks as part of their normal diet. For example, the per capita yearly consumption of fermented milks in Finland is 37.8 kg compared to the 4.2 kg in the UK. In Japan the probiotic approach has been adopted not only in the form of fermented milks but also as an additive to sweets, confectionery and soft drinks.

## Identification of Protective Flora

Many of the currently available probiotics contain lactic acid bacteria based on the concept first proposed by Metchnikoff. A more recent development has been attempts to establish the microbiological basis of the protective effect and to identify the micro-organisms involved. This phase in the development of probiotics dates from the mid 1950s and was stimulated by the work of Bohnhoff *et al.* (1954) and Freter (1955;1956) who showed that rodents could be rendered more susceptible to infection by administration of antibiotics. In his early experiments Freter (1956) was able to restore the resistance of the antibiotic treated guinea-pigs by colonising them with an antibiotic-resistant strain of *E. coli.* However, subsequent experiments showed that although this was the solution to this particular problem, it had no general application. Freter, himself, in later experiments used 95 different strains of bacteria to restore the normal degree of resistance (Freter and Adams 1972). These types of discrepancy between carefully designed and executed trials show how difficult it is to be sure that the experimental conditions are providing a suitable model which will yield results on which to base a general theory.

The way in which the indigenous gut microflora could antagonise the establishment of pathogens in the gut has also been studied in germ free animals. Early work by Maier *et al.* (1972) showed that *Shigella flexneri* rapidly colonised the gut of germfree mice reaching levels of $10^{10}$ cfu per gram within 24 h, whereas in conventional counterparts the count of *Sh. flexneri* declined to $10^3$ cfu per gram. Similar results were obtained with *Salmonella enteritidis* (Collins and Carter 1978) and *Campylobacter jejuni* (Jesudason *et al.* 1989).

A corollary of this finding that the gut flora protects against disease is that dosing with a faecal flora should restore resistance to an animal whose gut flora has been destabilised in some way. This sort of treatment has proved successful. Human patients suffering from pseudomembranous colitis caused by *Clostridium difficile* could be successfully treated with faecal enemas prepared from healthy adults (Eiseman *et al.* 1958; Bowden *et al.* 1981; Schwass *et al.* 1984). Faecal suspensions were also used to prevent colonisation of live chicken gut by salmonellae. Successful though these treatments were they were not the sort of therapy that could be recommended as routine practice; the possibility of introducing pathogens along with the protective flora was too great.

In order to find a more acceptable way of restoring the flora many groups have been engaged in research to identify the micro-organism which enable the conventional animal to resist disease. The general finding is that more than one strain of micro-organism is involved in reproducing the effect. This kind of phenomenon frustrates adequate experimentation. As Raiband (1992) has pointed out "...if 2 of 50 strains are involved in a synergistic action the number of combinations possible is as much as 1224; it is 19 600 if 3 of 50 strains are involved". It is not surprising, therefore, that in the gut which may contain up to 400 different strains, that sufficient trials have not been done to identify the organisms involved. The stage we have reached at the moment is the ability to reproduce, sometimes only partially, the effect with a large number of strains. This is well illustrated by the work done on competitive exclusion in the chicken (Mead and Impey 1987). Here the effect can be partially reproduced by 48 strains of micro-organism comprising *Lactobacillus acidophilus* (2 strains), *L. salivarius* (5), *L. fermentum* (1), *S. faecalis* subsp *liquefaciens* (1), *S. faecium* (1). *E. coli* (8), *Bacillus coagulans* (1), *Clostridium* sp. (5), *Cl. subterminale* (2), *Cl. tertium* (1), *Cl. cochlearium* (2), *Bacteroides hypermegas* (2), *Bact. vulgatus* (2), *Eubacterium* sp (2), *Bifidobacterium* sp. (2), *Peptostreptococcus* sp. (3), Anaerobic *Streptococcus* sp. (1), Anaerobic budding bacterium (3), and an anaerobic Gram-Positive curved rod (1).

Such results illustrate the problem of identifying which of the many components are essential for the manifestation of the effect; which strains on their own or combined with only one or two others would mimic the results obtained using complete microfloras or large numbers of strains derived therefrom. As stated earlier the number of combinations containing one, two or three strains selected from those available is vast and prohibits trials based on the logical methodology. Selection must be based on experience; a gut feeling.

Some information has been gained by exclusion experiments; for example, if the lactobacilli are omitted from the chicken collection the amount of protection is reduced (Mead and Impey 1987). We can, therefore, conclude that although lactobacilli on their own do not afford protection, they are an essential component of the protective flora. They may therefore be a useful probiotic supplement in the situation where most of the protective flora is already present but its effect is compromised by the absence of lactobacilli. It will also follow that lactobacilli will not be effective as a probiotic if other elements of the protective flora are missing; this would explain some of the variable effects obtained with lactobacillus-based probiotic supplements.

## Mechanism

Little is known about the way in which probiotic effects are achieved but our knowledge of the factors influencing gut microbial ecology can be used as a basis for intelligent speculation. Using this information it seems possible that probiotics are acting in several ways; the mechanism involved may be different for different types of probiotic.

### Direct Antagonism

One micro-organism may affect another by production of chemical compounds. *In vitro* tests have shown that organic acids as well as bacteriocins may be active in this respect. Results obtained *in vivo* suggest that volatile fatty acids (VFA) produced by anaerobic bacterial fermentations are likely to be important in the gut. Hentges and his group have related VFA levels to suppression in the intestine of *E. coli* and *Shigella flexneri* (Pongpech and Hentges 1989). Other workers (Freter 1992) have been unable to confirm the influence exerted by VFA and suggest that anaerobic bacteria exert suppression through a different mechanism.

## Competition for Nutrients

Gut contents are a rich source of microbial nutrients and it may seem unlikely that micro-organisms could not find sufficient food for growth. However, it should be remembered that in order to prevent growth the environment has only to be deficient in one essential nutrient. Even more, ability to rapidly utilise an energy source may reduce the log phase and make it impossible for the organism to resist the flushing effect exerted by peristalsis.

In the case of *Cl. difficile* it has been shown that nutrient deprivation is, at least in part, responsible for its failure to grow in the gut (Wilson and Perini 1988).

## Competition for Adhesion Receptors

There is no doubt that ability to adhere to the gut epithelium is an important colonisation factor. However, it is a very specific phenomenon and while it can be shown that non-pathogenic K88 *E. coli* can prevent adhesion by pathogenic K88 *E. coli*, it is difficult to envisage the specific blocking by lactobacilli of adhesive receptors for unrelated organisms such as *E. coli*. It may still be a factor but in a non-specific way i.e. by coating the gut epithelium with a continuous layer of lactobacilli the *E. coli* would be denied access to receptor sites. In the mouse stomach the colonisation of the squamous surface by lactobacilli prevents colonisation by yeast, but when the lactobacilli are removed by penicillin the yeast is able to attach and colonise (Savage 1969).

## Stimulation of Immunity

The ability of lactobacilli to influence the immune system is one of the more recent developments in this field (Bloksma *et al.* 1979). Subsequent research has shown that lactobacilli are capable of acting as immunomodulators by enhancing macrophage activity (Perdigon *et al.* 1986) increasing antibody levels (Bloksma *et al.* 1979; Yasui *et al.* 1989) inducing production of interferon (De Simone *et al.* 1986) and activating killer cells (Kato *et al.* 1984).

The scope and potential of probiotics is greatly increased by these findings. Although traditionally the effects of probiotics were thought to be limited to the gastrointestinal tract, it is now possible to envisage an influence on diseases in other areas of the body.

The work on immune stimulation by lactobacilli has been done almost entirely with laboratory animals and its relation to disease resistance in humans is not as yet clear.

# Effects on the Host

There is good evidence that the indigenous gut microflora provides protection against a wide range of infections. If that is so why should we need to resort to supplementation with probiotics? The answer lies in the way in which we maintain animals (including human babies) during the neonatal period. It is known that the composition of the gut microflora can be affected by diet, environmental conditions and emotional stress (see Tannock 1983). The conditions under which we rear young animals have all these elements and consequently the gut microflora may differ from that of an animal reared in the natural or the wild environment.

The incubator-reared chicken and the early weaned pig or calf are examples of rearing practices which restrict the contact that the offspring have with the mother and as a consequence the gut flora may be deficient in protective organisms and render the neonate more susceptible to disease.

In human babies the flora may vary even between different locations. For example the dominant *Bifidobacterium* sp. is different in different countries. In France the dominant bifidobacteria in breast-fed infants is *B. bifidum* (Neut *et al.* 1980), in Japan it is *B. breve* (Yuhara *et al.* 1983) and in Italy, *B. infantis* (Biovati *et al.* 1984). There are even detectable differences within the same country; Poupard *et al.* (1973) found it easier to detect bifidobacteria in babies from small suburban hospitals than in those from large urban hospitals and Lundequist *et al.* (1985) found differences between wards in the same hospital. The different environmental conditions and nursing practices may account for these differences in the gut microflora of babies.

Maternal contact is also a factor which can affect the gut flora. Babies delivered by caesarian section into incubators had a reduced incidence of lactobacilli in their faeces (Hall *et al.* 1990).

It is in situations like this that probiotic supplementation has its greatest potential. The restoration of the microflora to its natural composition is likely to have a major effect on the resistance of the neonate to disease.

This applies also in the adult whose gut flora has been deranged by factors such as oral antibiotic treatment and unaccustomed dietary habits such as are experienced during travel abroad. One result of the first stress factor is pseudomembranous colitis caused by *Cl. difficile*; this condition has been amenable to treatment with an avirulant strain of *Cl. difficile* (Borriello and Barclay 1985) and a *Lactobacillus* sp. (Gorbach *et al.* 1987).

Another good example of the effect that probiotics can have on human adults is their use for alleviation of lactose deficiency. The amount of lactose which causes symptoms when given in milk can be utilised without symptoms if given in yoghurt (Kolars *et al.* 1984).

Probiotic supplementation can affect the metabolic activity of the gut flora. Administration of *L. acidophilus* to human patients reduced the activity of enzymes which convert pro-carcinogens to proximal carcinogens (Goldin and Gorbach 1984). The suggestion is that this may be a treatment which would prevent or delay the onset of cancer and there is some supporting evidence for this in studies in which *L. acidophilus* prolonged the induction of the latent period in animals give the chemical carcinogen 1,2-dimethylhydrazine (Goldin and Gorbach 1980).

There are also reports of positive effects on the growth rate of animals, improved milk yield of cows and increased egg production by chickens. The relevance of the effects to the human being is at the moment not obvious.

The effects listed above are supported by experimental evidence which is often statistically significant. However, the results are variable and therefore difficult to summarise. The reasons for this variability have been discussed elsewhere (Fuller 1990). Briefly, they include factors such as

a variation of the indigenous microflora; if the stressing micro-organism is not present there is no scope for improvement;

b the strain of probiotic organism used; variations between strains of the same species occur;

c viability of the preparation; this is not always checked and low viability may account for some negative results;
d age of the animal; different results may be obtained with suckling and weaned animals;
e methods of production; the way in which the probiotic organism is grown and harvested may affect its behaviour in the intestinal tract.

With all these possible variable factors it is not surprising that probiotics appear to give inconsistent results. But we should take heart from the fact that under the right conditions and using the right organism, produced in the right way, significant responses can be obtained.

# Future Developments

The ideal probiotic would be one which could establish itself permanently in the intestine and produce its active agents *in situ*.

In order to select or genetically design such an organism we must know:

a the factors which determine colonisation of the gut, and
b the biochemical basis of the probiotic effect.

While we know something about the former, with the exception of the aleviation of lactose deficiency, we know almost nothing about the latter. The elucidation of the mode of action is absolutely crucial to the further rational development of improved probiotic preparations. It is only then that we can use biochemical characteristics to recognise potentially useful probiotic organisms in the laboratory. Without this sort of marker we are reduced to clinical trials and animal field trials which are time consuming and expensive.

The recognition of the mechanism would also allow us, depending on how it worked, to genetically modify strains to augment the effect or to transfer it from a strain which colonised poorly to one which colonised well.

Fractionation of the probiotic agent might also drive the development in another direction. If a stable compound responsible for the probiotic effect could be extracted it would remove the problem of having to sustain viability over long periods and lead to a second generation of non-viable probiotics. This type of product would also be amenable to improvement by genetic engineering without the problem of release of genetically modified micro-organisms into the environment. There are at the moment some non-viable microbial stimulants being studied but these are usually obtained from non-microbial sources.

The recent developments in immunomodulation could also be capitalised on by introducing proactive epitopes from pathogens into lactobacilli such as *L. casei*. In this context also the relationship between translocation from the gut to immune response could be studied.

The probiotic concept is becoming more important in clinical practice. While they may never replace antibiotics as therapeutic agents they may well have a future as prophylactic treatments, especially for intestinal diseases.

While there is no doubt that probiotics are effective under the right experimental conditions, in practice they are not always effective. What is essential to the future development and understanding of probiotics is a detailed knowledge of their mode of action. When this is available we will be able to select, develop and even design more effective strains.

# References


Barrow PA, Tucker JF, Simpson JM (1987) Inhibition of colonisation of the chicken alimentary tract with *Salmonella typhinurium* by Gram-negative facultatively anaerobic bacteria J Hyg 98:311-322

Bassler A, Lutz JR (1922) *Bacillus acidophilus*: its very limited value in intestinal disorders. J Am Med Ass 79:607-608

Benno Y, Mitsuoka J (1986) Development of the intestinal microflora in humans and animals Bifid. Microflora 5:13-25

Bibel DJ (1988) Elie Metchnikoff's Bacillus of Long Life. ASM News 54:661-665

Biovati B, Castagnoli P, Crociani F, Trovatelli LD (1984) Species of *Biofidobacterium* in the faeces of infants Microbiologica 7:341-345

Bloksma N, de Heer E, van Dijk M, Willers M (1979) Adjuvanicity of lactobacilli I differential effects of viable and killed cells. Clin Exp Immunol 37:267-375

Bohnhoff M, Drake BL, Miller CP (1954) Effect of streptomycin in susceptibility of intestinal tract to experimental *Salmonella* infection Proc Soc Exp Biol Med 86:132-137

Borriello SP, Barclay FE (1985) Protection of hamsters against *Clostridium difficile* ileocaecitis by prior colonisation with non-pathogenic strains. J Med Microbiol 19:339-350

Bowden TA, Mansberger AR, Lykins LE (1981) Pseudomembranous enterocolitis: mechianism for restoring floral homeostasis. Am Surg 47: 178-183

Bullen CL, Tearle PV, Willis AT (1975) Bifidobacteria in the intestinal tract of infants: an *in vivo* study. J Med Microbiol 9:325-333

Collins FM, Carter PB (1978) Growth of salmonellae in orally infected germfree mice. Infect Immun 21:41-47

De Simone C, Bianchi-Salvadori B, Negri R, Ferrazzi M, Baldinelli L, Besely R (1986) The adjuvant effect of yogurt on production of gamma-interferon by ConA stimulated human peripheral blood lymphocytes. Nutr Rpts Int 3:419-431

Duval-Iflah Y, Ouriet MF, Moreau C, Daniel N, Gabilan JC, Raiband P (1982) Implantation précoce d'une southe de *Escherichia coli* dans l'intestin de nouveau-nés humains: effet de barrière vis-a-vis des souches de *E. coli* antibiorésistantes. Ann Microbiol (Inst Paster) 133A:393-408

Eiseman B, Silem W, Bascomb WS, Kanvor AJ (1958) Fecal enema as an adjunct in the treatment of pseudomembranous enterocolitis. Surgery 44:854-858

Florey HW (1946) The use of micro-organisms for therapeutic purposes. Yale J Biol Med 19:101-117

Freter R (1955) The faecal enteric cholera infection in the guinea pig chieved by inhibition of normal enteric flora. J Infect Dis 97:57-64

Freter R (1956) Experimental enteric *Shigella* and *Vibrio* infection in mice and guinea pigs. J Exp Med 104:411-418

Freter R (1992) Factors affecting the microecology of the gut. In: Fuller R (ed) Probiotics. The scientific basis, Chapman and Hall, London pp 111-144

Freter R, Abrams GD (1972) Function of various intestinal bacteria in converting germfree mice to the normal state. Infect Immun 6:119-126

Frost WD, Hankinson H (1931) Lactobacillus acidophilus. An annotated bibiography. Davis-Greene Corporation, Milton Wisconsin

Fuller R (1989) Probiotics in man and animals. J Appl Bact 66:365-378

Fuller R (1990) Probiotics in Agriculture Ag Biotech News 2:217-220

Goldin BR, Gorbach SL (1980) Effect of *Lactobacillus acidophilus* dietary supplementation on 1,2-dimethylhydrazine dihydrochloride-induced intestinal cancer in rats. J Natl Cancer Inst 64:263-265

Goldin BR, Gorbach SL (1984) The effect of milk and lactobacillus feeding on human intestinal bacterial enzyme activity. Am J Clin Nutr 39:756-761

Gorbach SL, Chang TW, Goldin BR (1987) Successful treatment of relapsing *Clostridium difficile* colitis with *Lactobacillus* GG. Lancet ii,1519

Hall MA, Cole CB, Smith SL, Fuller R, Rolles CJ (1990) Factors influencing the presence of faecal lactobacilli in early infancy. Arch Dis Childhood 65:185-188

Hughens HV (1925) Colonic irrigation and *B. acidophilus* diarrhœa. US Navy Med Bull 22:691-692

Hull TG, Rettger LF (1917) The influence of milk and carbohydrate feeding on the character of the intestinal flora. J Bact 2:47-71

Jesudason MV, Hentges DJ, Pongpech P (1989) Colonisation of Mice by *Campylobacter jejuni*. Infect Immun 57:2279-2282

Kato I, Yokokura T, Mutai M (1984) Augmentation of mouse natural killer cell activity by *Lactobacillus casei* and its surface antigens. Microbiol Immunol 28:2099-217

Kolars JC, Levitt MD, Aouji M, Savaino DA (1984) Yogurt- an autodigesting source of lactose. New Engl J Med 310:1-3

Kopeloff N (1923) Is Bacillus acidophilus therapy a strictly bacteriological phenomenon? Proc Soc Exp Biol Med 10:123-124

Lilley DM, Stillwell RJ (1965) Probiotics: growth promoting factors produced by micro-organisms. Science 147:747-748

Lloyd-Evans LPM (1989) Probiotics PJB Publications Ltd

Lundequist B, Nord CE, Winberg J (1985) The composition of the faecal microflora in breast-fed and bottle-fed infants from birth to eight weeks. Acta Paediatr Scand 74:45-51

Maier BR, Onderdonk AB, Baskett RC, Hentges DJ (1972) *Shigella*, indigenous flora interactions in mice. Amer J Clin Nutr 25:1433-1440

Mead GC, Impey CS (1987) The present status of the Nurmi concept for reducing carriage of food-poisoning salmoellae and other pathogens in live poultry. In: Smulders FJM (ed) Elimination of pathogenic organisms from meat and poultry Elsevier Ansterdam pp57-77

Metchnikoff E (1907a) Essais Optimistes. A Maloine, Paris

Metchnikoff E (1907b) The Prolongation of Life, Optimistic Studies, Heinemann, London

Neut C, Romond C, Beerens HA (1980) A contribution to the study of the distribution of *Bifidobacterium* species in the faecal flora of breast-fed and bottle-fed babies. Reprod Nutr Dev 20:1679-1684

Orla-Jensen S (1912) Maelkeribakterologi. Schønbergske forlag, Copenhagen

Parker RB (1974) Probiotics, the other half of the antibiotics story. Anim Nutr Hlth 29:4-8

Perdigon G, Nader de Marcias ME, Alvarez S, Oliver G, de Ruiz Holgado AAP (1986) Effect of perorally administered lactobacilli on macrophage activation in mice. Infect Immun 53:404-410

Pongpech P, Hentges DJ (1989) Inhibition of *Shigella flexneri* and enterotoxigenic *Escherichia coli* by volatile fatty acids in mice. Microbiol Ecol Hlth Dis 2:153-161

Poupard JA, Husain I, Norris RF (1973) Biology of Biffidobacteria Bacteriol. Rev 37:136-165

Raiband P (1992) Bacterial interactions in the gut In: Fuller R (ed) Probiotics. The scientific basis. Chapman and Hall London pp 9-28

Rettger LF, Chaplin HA (1921) Therapeutic application of *Bacillus acidophilus* Proc Soc Exp Biol Med 19:72-76

Robertson B, Roerstad G, Engstad R, Raa J (1990) Enhancement of non-specific disease resistance in Atlantic salmon *Salmo salar* L., by glucan from *Saccharomyces cerevisiae* cell walls. J Fish Dis 13:391-400

Rotch TM, Kendall AI (1911) A preparatory study of the Bacillus acidophilus in regard to its possible therapeutic use. Am J Dis Child 2:30-38

Savage DC (1969) Microbiol interference between indigenous Yeast and Lactobacilli in the rodent stomach. J Bacteriol 98:1278-1283

Schwass A, Sjolin S. Trottestam U, Aransson B (1984) Relapsing *Clostridium difficile* enterocolitis cured by rectal infusion of normal faeces. Scand J Infect Dis 16:211-215

Sperti GS (1971) Probiotics Avi Publishing Co Inc, Westpoint Connecticut

Tannock GW (1983) Effect of dietary and environmental stress on the gastrointestinal microbiota. In: Hentges DJ (ed) Human Intestinal Microflora in Health and disease. Academic Press, New York pp 517-539

Tissier H (1905) Repartition des microbes dans l'intestin du nourisson Ann Inst Past 19:109-115

Wilson KH, Perini F (1988) Role of competition for nutrients in suppression of *Clostridium difficile* by the colonic microflora. Infect Immun 56:2610-2614

Yahura T, Isojima S, Tsuchiya F, Mitsuoka T (1983) On the intestinal flora of bottle-fed infants. Bifid Microflora 2:33-39

Yasui H, Mike A Ohwaki M (1989) Immunogenicity of *Bifidobacterium breve* and change in antibody production in Peyer's patches after aral administration. J Dairy Sci 72:30-35

Chapter 5

# Strategies for the Isolation and Characterisation of Functional Probiotics

P. L. Conway and A. Henriksson

## Introduction

### General

The age-old concept of using a range of specific microorganisms for therapeutic purposes in man has been applied to clinical conditions involving many organs e.g. the gastrointestinal tract, oro-pharyngeal region, pulmonary disturbances, vaginal and urinary tract infections. In a review of the use of microorganisms in therapy, Florey (1946) cites the work of Cantani published in 1885 as the first reported human study. In this case, an organism considered to be harmless was administered to a tuberculosis patient. Metchnikoff (1907) proposed a prophylactic benefit from both nurturing the indigenous lactobacilli of the gastrointestinal tract by attention to the diet, as well as from ingesting milk fermented with lactobacilli. Subsequently, most attention has focussed on lactic acid bacteria as being potentially beneficial.

The validity of the hypothesis that microorganisms and in particular lactic acid bacteria can be used therapeutically and prophylactically has been extensively investigated over the years. Unfortunately, the studies of yester-year were often not adequately designed to conclusively prove or disprove this use of microorganisms. In the light of the continued controversy, the numerous positive trends and need for alternative treatments, a new generation of such preparations is emerging. The hallmark of the approach used today is that considerable attention is paid to both the selection of the specific microorganism and the experimental design for evaluating the function. Strategies being used for developing this new generation of preparations are presented here. Initially, we have defined our terminology, and then discuss generally the principles used for isolation and characterisation of probiotic microorganisms. Examples of our application of these concepts using lactobacilli within the gastrointestinal tract are presented. In view of the recent reviews on the topic of the health and nutritional benefits of lactic acid bacteria (e.g. Gilliland 1989, 1990, Tannock 1990, Juven *et al.* 1991) and probiotics (Fuller 1992), no attempt is made to comprehensively cover the literature in this area but rather examples of the various aspects are cited and discussed in brief.

### Terminology

Although the use of the probiotic approach dates back to before the turn of the century, the term was only coined in 1965 by Lilly and Stillwell to describe the growth promoting effects of one microorganism on another. Subsequently, various interpretations of the term have been presented. Throughout this chapter, the term probiotic will be utilised to refer to all live microbial preparations which mediate (demonstrable) beneficial effects for the host, in this case humans. We favour insertion of the term "demonstrable" for probiotics of the future, although recognise that many of the preparations to date have not been evaluated on these grounds. By maintaining this relatively broad definition, the term is not limited to the gastrointestinal tract. We also favour "microbial" rather than bacterial so as to include yeasts, which is not the case for some interpretations, e.g. bacterial interference. Although some workers use the term probiotic to include "substances", presumably of microbial origin, we restrict our usage to microbes only. Furthermore, our interpretation of the term encompasses preparations which can induce physiological and biochemical responses in the host, rather than being restricted to preparations which reduce infection.

## Probiotics of Yesteryear

In order to develop strategies for the isolation of new functional probiotics, considerable information can be gleaned by evaluating the preparations of yester-year. The term functional is stressed because over the decades, many unanswered questions arise as to the effectiveness of probiotics, in particular those based on lactobacilli.

### Lactic Acid Bacteria Based Preparations for the Digestive Tract

Very many beneficial roles and effects have been reported for lactic acid bacteria, and particularly lactobacilli, within the gastrointestinal tract. These can be broadly divided into physiological, biochemical and antimicrobial effects, many of which can be attributed to effects on the indigenous and/or transient microbiota. Reported physiological effects include stimulation of the immune system (Gerritse *et al.* 1990), decreased incidence of tumours and improved gastrointestinal motility (Friend *et al.* 1982). The biochemical effects that have been reported for lactobacilli encompass decreased faecal carcinogen and mutagen levels, assimilation and reduction of cholesterol, deconjugation of bile acids (Gilliland and Speck 1977), improved nutritional status of foods, improved tolerance of lactose, metabolism of drugs and inactivation of mutanogenic factors (Goldin 1986,). The parameters directly attributable to antimicrobial effects include: contribution to the establishment and stabilisation of the gut microbiota (Morotomi *et al.* 1975, Muralidhara *et al.* 1977)), protection against pathogen colonisation and reduced incidence of bacterial diarrhoea (reviewed by Fernandez *et al.* 1987). This long and varied list of beneficial roles of lactic acid bacteria must evoke scepticism in the minds of many. In fact, the role of lactobacilli is one of the most controversial subjects of gut ecology. Doubts have been strengthened by publication of apparently contradictory results, non-conclusive reports and statistically non-significant results (as previously reviewed Conway 1989). In retrospect, many of the inconsistencies between studies of yesteryear can be attributed to:

i species and strain differences;
ii lack of viability of the preparation;
iii subject variability.

## Species and Strain Differences

It is paramount in this issue to recognise that not all species of lactic acid bacteria will have similar characteristics and that strains of the same species will also vary. Furthermore, because a particular strain or strains produce a good fermented milk product, this does not automatically meant it will function well as a probiotic. Any preparation is only as effective in eliciting a specific response as is the particular strain of organism used. Ignorance of this fact leads to unjustified general criticisms of probiotics because one specific preparation is ineffective. Most recent reviews of the use of lactic acid bacteria as dietary adjuncts address this issue (e.g. Gilliland 1989, 1990).

## Viability

Another category to consider is the viability of the preparation. Lactobacilli are particularly sensitive to freeze drying procedures, failing to thrive after rehydration if careful attention is not paid to this process. Many of the earlier freeze dried preparations contained extremely low levels or variable levels of viable cells (Robins-Browne and Levine 1981).

## Subject Variability

Environmental factors influence the hosts susceptibility to infection and therefore the action of the probiotic directed against the infection. For example, it is well established that stress can cause alterations in the gastrointestinal microbiota (reviewed Tannock 1983) which in turn has ramifications for susceptibility to infection. Genetic and physiological variations in subjects will also influence the impact of an administered probiotic preparation. Consequently, insufficient numbers of subjects can also lead to non-conclusive results. Recently, Asby *et al.* (1989) used regression analysis of data to remove many of the variables associated with early lamb growth and could show statistically significant effects of probiotic dosage.

## Other Microbial Preparations

Generally, the use of microbes other than lactic acid bacteria tends to be less clouded by dogma of by-gone years and studies. Very many other microbes have been used as probiotics for a range of organs. The rational for the selection of the particular strain has often been more specific than that used for lactic acid bacteria. That is, instead of attempting to improve the microbial balance which is not readily measurable, specific conditions have been targeted. For example, in 1956 Sears *et al.* introduced an avirulent *Escherichia coli* into the intestine of a dog and thereby protected the host from a challenge from a virulent strain of *E. coli.* Shortly after, the mildly virulent *Staphylococcus aureus* strain 502A was used to successfully block colonisation of the pharynx of neonates by a very virulent strain which would otherwise have caused an epidemic in the hospital nursery (Shinefield *et al.* 1963). In order to protect the host from *Streptococcus mutans* induced dental caries, workers (e.g. Kurasz *et al.* 1986; Willcox and Drucker 1988) are developing preparations based on *Streptococcus spp* which do not mediate caries.

# The New Generation Probiotics

With the increasing development of antibiotic resistance, restricted usage of antibiotics for farmed animals and awareness of clinical conditions which do not respond to antibiotics, there is considerable need for alternatives. A new generation of probiotics is emerging. The hallmark of these new preparations is that the specificity and function are defined. In particular, the target is identified, the strain specifically selected, preparatory conditions designed to ensure stability of desirable characteristics, and the status of the target ecosystem considered. For example, lactobacillus strains are now being selected on these grounds rather than the narrower view that it is only necessary to maintain high numbers of the microbe in situ.

## Identification of the Target

A clear definition of the target will allow development of a specifically directed preparation. General terms such as "beneficial to the host", "improve intestines" and "intestinal balance" will be replaced by more scientifically exact statements specifying the site of action of the probiotic microbe. For example, a probiotic preparation designed to protect the host from traveller's diarrhoea will not automatically be assumed to also have an effect against mutagens or to reduce cholesterol. One can envisage that ultimately probiotics for travellers diarrhoea may even specify the geographical area of recommended usage. This will be feasible by initially studying the major causes of diarrhoea among travellers in a specific geographic location and then selecting strains with biological activity against that causative agent, as outlined below. An alternative could be to have cocktails of various strains such that a broad spectrum of biological activity is obtained. Black *et al.* (1989) successfully reduced diarrhoea in travellers to

Egypt from 71% to 43% by dosing a mixture of *Lactobacillus acidophilus*, *L. bulgarius*, *Bifidobacterium bifidus* and *Streptococcus thermophilus*. A cautionary note should also be made that with many conditions e.g. travellers diarrhoea, there are most probably numerous causative agents. One of the primary causes of travellers diarrhoea is the enterotoxigenic *E. coli* (Gorbach *et al.* 1975). In any specific tourist resort, there are most likely several typical food-born pathogens as well as the cases of non-bacterial diarrhoea induced by major changes in diet, a factor know to cause gastrointestinal disturbances (Winitz *et al.* 1970, Hentges 1983). Such a situation could explain why conflicting results can be obtained for the same preparation when evaluated in two different locations. For example, Oksanen *et al.* (1990) treated 50% of travellers at two different locations with *Lactobacillus* GG and showed a marked effect on the incidence of diarrhoea in only one destination.

Probiotics have been proposed as a treatment for constipation in the elderly (Alm *et al.* 1983), however, the causative agent is not so easily established. It has been suggested that decreased fibre intake is a major contributing factor and in this case it would not be possible to specifically select a probiotic microbe which is biologically active against this factor. It is plausible that a probiotic could be targeted, however, towards reversing the physiological state induced by the lowered fibre intake. Borody *et al.* (1989) recently presented evidence that an infective aetiological agent may be involved in chronic constipation as well as in inflammatory bowel disease and irritable bowel syndrome. These workers suggest that chronic diarrhoea, ulcerative colitis or Crohn's disease may also be targetable by microbial therapy. Identification of the causative agent(s) would allow selection of specific probiotic strains inhibitory to this agent.

While identification of the target may be feasible in cases of infection, it may be unrealistic to suggest identifying the causative agent in other cases, e.g. for tumours. In such cases it may be more appropriate to indirectly affect the target. Although the causative agents of tumours are most often not identified, it has recently become apparent that there is a correlation between the presence of *Helicobacter pylori* chronic infection dating from childhood and the incidence of stomach cancer. Consequently, a probiotic directed against *H. pylori*, such as the *L. acidophilus* strain shown to inhibit growth of *H. pylori* (Bhatia *et al.* 1989) in the long term, may contribute to decreased incidence of stomach cancer. To date, probiotic studies directed towards reducing the incidence of tumours have concentrated on indirect targeting by enhancing the immune response (Kato *et al.* 1983), reducing the presence of colonic enzymes which can convert procarcinogens to carcinogens (Goldin and Gorbach 1980 1984) and by reducing the levels of faecal mutagens (Lidbeck *et al.* 1991; Morotomi and Mutai 1986, Zhang *et al.* 1990).

## Strain Selection

Selection of the strain to be used is by far the most important aspect in the strategy of producing functional probiotics as acknowledged by Gilliland back in 1979 and more recently discussed by many (e.g. Goldin and Gorbach 1987; Gilliland and Walker 1990; Jonsson and Conway 1992). This is therefore discussed in brief here and the major issues expanded upon in the following sections. It is now generally accepted that the selected strains must originate from healthy subjects and from the intended organ for usage. It can not be assumed however, that all indigenous microbes would be suitable for selection. Lactic acid bacteria are still the most commonly selected strains. Because the entire indigenous microbiota may have numerous beneficial effects on the host, colonic microbiota as a whole, instead of isolated strains, has also been used to successfully reverse symptoms of constipation and irritable bowel syndrome in humans (Borody *et al.* 1989, Andrews *et al.* 1992). Tvede and Rask-Madison (1989) have cultured ten different indigenous colonic microbes and reversed symptoms in patients with *Clostridium difficile* associated diarrhoea. Similarly Rolfe *et al.* (1981) reduced *C. difficile* associated diarrhoea using a mixture of 23 indigenous anaerobic and aerobic bacteria.

Furthermore, the selected strain must have the capacity to be viable and occur in significant numbers within the target organ and this is frequently discussed in terms of the capacity of the strain to colonise the site. These parameters have evolved from the recognition that the indigenous microbes contribute to developing and stabilising the ecosystem and thereby protect the host from invading pathogens and toxins. Another major consideration is that the selected strain must be biologically active against the target pathogenic factor. Considerable emphasis is now placed on the issues of colonisation and biological activity as outlined below.

**Table 5.1.** Effect of sub-culturing in MRS broth on adhesive character of a number of lactobacillus strains

| | No. of transfers in MRS broth | | | |
|---|---|---|---|---|
| Lactobacillus | 5 | 10 | 30 | 60 |
| strain 737 | + | ± | - | - |
| strain C39 | + | + | - | - |
| strain 152 | + | + | + | + |

## Stability of Characteristics

It is not only important to ensure viability of the probiotic strain but one must also ensure that the desired characteristics of the strain are stable. Clements *et al.* (1983) report lot to lot variation in the effectiveness of preparations of lactobacilli in reducing the volume and duration of neomycin associated diarrhoea. It has also been reported that freeze drying can be detrimental to surface proteins of lactobacilli and this could affect the adhesive nature of the strains (Bhowmilk *et al.* 1985). We have noted considerable variation in the stability of adhesive capacity during sub-culturing. As shown in Table 5.1, some strains such as *L. fermentum* 737 rapidly become non-adhesive while others such as *L. crispatus* strain 152 maintained the characteristic after sub-culturing. Considerably variety in response to subculturing 20 times has been reported for a range of lactobacilli when examined in terms of the surface hydrophobicity of the bacterial cell (Reid *et al.* 1992). Sub-culturing can also induce *Lactobacillus* strains to alter their colony morphology and this is accompanied by altered characteristics of the strain (Klaenhammer and Kleeman 1981). Instability is not just a phenomenon for lactobacilli. Natural isolates of *Escherichia coli* have been shown to rapidly change when grown in continuous culture and become consistent with laboratory strains when assessed by their ribosomal efficiency and numbers (Mikkola and Kurland 1991). Questions can, therefore, be raised about the stability of avirulent strains used to outcompete the virulent pathogenic form e.g. as shown for *Staphylococcus aureus*

(Shinefield *et al.* 1963) and *E. coli* strains not expressing colonisation or resistance factors (eg Duval-Iflah *et al.* 1982). No risks would be associated with the use of a strain which is avirulent because of the loss of genes encoding for virulence factors, however, if the strain retains a gene encoding for a virulence factor, the expression could be triggered by *in vivo* conditions. Transfer of plasmids coding for virulence factors among enteric organisms in the gut is well documented (Smith and Linggood 1971, Ørskov *et al.* 1975) and may represent a considerable hurdle to the clinical use of avirulent strains.

## Status of the Ecosystem

The status of the target ecosystem for the probiotic will play a significant role in the effectiveness of the microbe. For example, the gastrointestinal tract is a complex and dynamic ecosystem balanced delicately by contributions from the host physiology, the diet and the indigenous microbiota (Savage 1972; Hentges 1983). Alterations in diet and nutrient deprivation have been shown to alter the profile of lumenal and surface associated bacterial populations. Lactobacilli and enterobacteria such as *E. coli*, emerge as indicators of such stress with the lactobacilli invariably being sensitive and the enteric bacteria proliferating(reviewed Tannock 1983). For example, lactobacilli colonising the epithelial mucosa were lost and the large intestinal aerobic population shifted from containing about 1% enteric bacteria to being virtually 100% enteric bacteria when rodents were deprived of nutrients for 24 hours (Conway *et al.* 1986).

An understanding of the interactions within the ecosystem will allow better anticipation of the effectiveness of an administered probiotic strain. An example of the

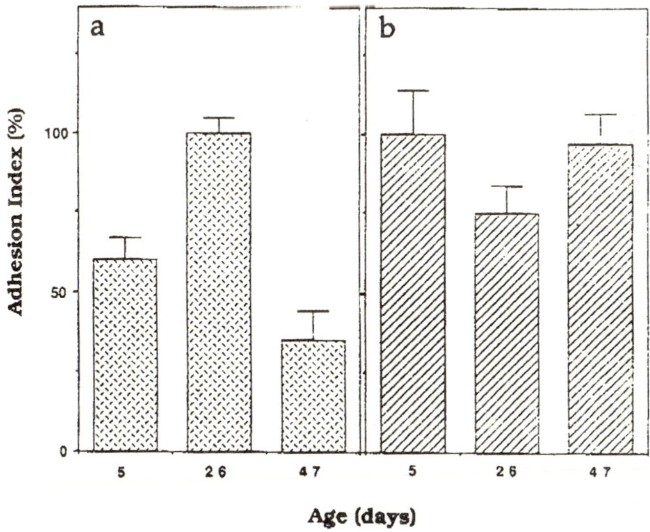


**Fig. 5 1.** Adhesion to porcine gastrointestinal epithelial mucosa of piglets 5, 26 and 47 days old: (**a**) adhesion of *E. coli* K88 cells to ileal mucus; (**b**) adhesion of *Lactobacillus* strain 152 to squamous gastric tissue (after Blomberg and Conway 1989).

host contribution to the colonisation potential of an ingested probiotic strain can be seen in the results presented in Figure 5.1 (Blomberg and Conway 1989). In this experiment, we studied the adhesion (as a measure of colonisation potential) of a lactobacillus strain of porcine gastric origin and pathogenic *E. coli* K88 cells to epithelial mucosa from the digestive tract of 5, 26 and 45 day old piglets. Of these three age groups, 26 day old piglets are much more sensitive to *E. coli* K88 induced diarrhoea. *Lactobacillus* adhesion to mucosal epithelium from 26 day old piglets was suppressed, while adhesion of *E. coli* K88 was enhanced relative to the 5 and 45 day mucosal samples. The age related changes in the epithelial mucosa influence the susceptibility of the host to colonisation by potentially beneficial probiotic lactobacilli and the undesirable pathogenic *E. coli* K88.

## Colonisation Capacity of Probiotics

The term colonisation when referred to probiotics describes the situation when the administered microorganism establishes itself and continues to multiply within the organ. *In vivo* colonisation is measured by detection of the strain after administration. Some studies use this term to describe the detection of the strain during administration (Saxelin *et al.* 1991) and to be strictly correct this could only be referred to as transient colonisation. In this instance, colonisation is being used to refer to the fact that viable cells can be detected in situ. Distinction should be made between this use of the term "colonisation" and that referring to establishment or persistence of the strain. That is, detection of the strain for a sufficiently long time after cessation of dosage to ensure that the strain has been able to grow and withstand being removed from the system.

Colonisation of the target organ by the probiotic strain is most often discussed in terms of the capacity of the strain to: (a) adhere to mucosal surfaces and thereby resist being washed out of the system; (b) produce metabolites inhibitory or antagonistic to the indigenous microbiota which protect the host from colonisation by extraneous microbes; (c) survive the rigours of transit through the tract to the possible site of colonisation.

### Adhesion to Epithelial Mucosa

It is frequently stated now that for microbes to colonise an ecosystem or organ such as the digestive tract, they must be able to adhere to the epithelial mucosa in order to avoid being flushed out of the system. In fact, published studies are appearing in which it is reported that the strain under investigation is obviously a good candidate for use as a probiotic because it can adhere *in vitro*. As discussed below, this is an over-simplification of the situation. The hypothesis is based on the fact that pathogens are very frequently equipped with surface located virulence factors which mediate adhesion to epithelial mucosa and thereby facilitate colonisation (reviewed by Gaastra and de Graaf 1982). It is therefore often assumed that the same must apply to the components of the indigenous microbiota that are being selected as potential probiotics. Except for adhesion of lactobacilli to non-secretory epithelia in animals (Fuller 1973; Barrow *et al.* 1980; Lin and Savage 1984), evidence of constituents of the indigenous microbiota adhering to secretory epithelia in the digestive tract is sparse.

The mucus overlying the epithelial cells has been shown to contain receptors for pathogens (reviewed e.g. Karlsson 1989). For example, piglet ileal mucus contains receptors for enteropathogenic *E. coli* K88 (Blomberg and Conway 1989) and this mucus has been proposed as a site of colonisation, independent of adhesion (Conway *et al.* 1990, 1991). If a strain can not adhere to the epithelial cell surface, theoretically it could still colonise the overlying mucus or the lumen if its doubling rate is less than the flow rate of the system, thus avoiding being washed out. This is equally valid for non-pathogenic potential probiotic microbes.

Mucosal adhesion can be assessed in an *in vitro* adhesion assay, however the results must be interpreted with an understanding of mechanisms of adhesion and always in relation to the appropriate controls. Microorganisms can adhere specifically and non-specifically to surfaces (Rutter *et al.* 1984). Specific adhesion is described as the situation occurring when a component on the bacterial surface binds to a particular receptor on the mucosal surface in what can be described as a lock and key interaction. Non-specific adhesion is a more general phenomenon mediated by, for example, hydrophobic or electrostatic interaction. Consequently, bacteria can adhere non-specifically to a range of surfaces. For many of the enteropathogenic *Escherichia coli*, specific adhesion is mediated by adhesive proteinaceous fimbriae e.g. K88, K99, CFAI, that have been studied *in vitro* using antibodies to the purified adhesin as well as isogenic mutants which lack or carry the genes encoding for the adhesin.

To illustrate non-specific and specific adhesion, and provide evidence that demonstrable *in vitro* adhesion alone does not guarantee that the strain can adhere *in vivo*, results are presented in Table 5.2 for K88 fimbriae mediated adhesion. *E. coli* K12 cells carrying the plasmid encoding for K88 fimbriae, adhere approximately 100-fold more to piglet ileal mucus than the isogenic parent *E. coli* lacking the plasmid encoding for the K88 fimbriae, i.e. log 7.34 and log 5.60, respectively. This adhesion was inhibited using a monoclonal antibody specific for the K88 fimbriae, thus yielding approximately log 4.93 bacteria per test which is consistent with the number of K12 cells which adhere to mucus. This log 5 level of binding is similar to that obtained for adhesion of either K12 or K12+K88 to bovine serum albumin (BSA) which is used here as an indication of the extent of non-specific binding to proteins *per se*. *In vivo*, *E. coli* K88 bearing cells can colonise, while *E. coli* K12 cannot.

**Table 5.2.** Adhesion of isogenic *E. coli* strains K12 and K12 (K88) to pig small intestinal mucus in the absence or presence of monoclonal antibodies specific for the K88 fimbriae

| | Log number of bacteria adhering[c] | |
|---|---|---|
| Surface | *E. coli* K12 | *E. coli* K12 + K88[b] |
| BSA[a] | 5.70 | 6.49 |
| Mucus | 5.60 | 7.34 |
| Mucus + Antibody | 5.20 | 4.93 |

[a] BSA = Bovine serum albumin
[b] carries the plasmid containing the gene encoding for K88 fimbriae
[c] Results expressed as log number of bacteria adhering per unit area

This type of illustration is more difficult to exemplify without identifying and purifying the adhesin or without mutants varying in adhesive characteristics. Published studies on identified adhesins of non-pathogenic microbes and potential probiotic strains is sparse. Kleeman and Klaenhammer (1982) addressed the issue of different types of binding by showing calcium dependent and independent adhesion of *Lactobacillus* strains to human fetal intestinal cells. The adhesion of these

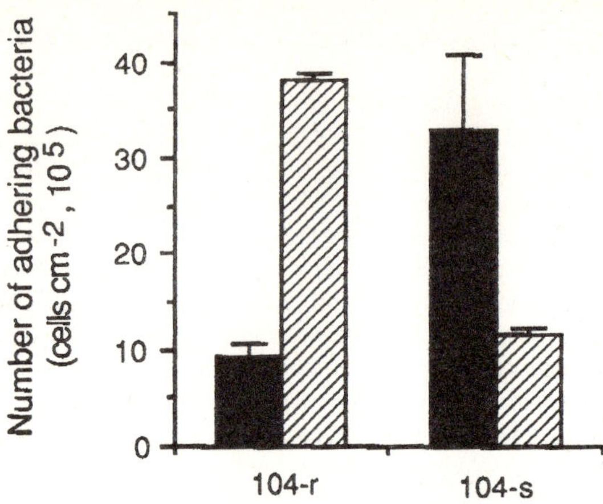


**Fig. 5.2.** Adhesion of *L. fermentum* 104-S and 104-R to polystyrene (shaded bars) and squamous gastric epithelium (solid bars). Results expressed as the number of cells adhering per square centimeter (after Henriksson *et al.* 1991).

*Lactobacillus* strains was further tested using fresh ileal cells of human and porcine origin (Conway *et al.* 1987). In this study, the same strain adhered well to both human and porcine cells, suggestive that this strain adhered non-specifically to secretory epithelial cells. It is important to stress that this does not mean that adhesion of all lactobacillus strains to intestinal epithelial cells is host non-specific.

Host specific adhesion of lactobacilli to non-secreting squamous tissue in various animals has been demonstrated *in vitro* (Fuller 1973; Barrow *et al.* 1980) and *in vivo* (Tannock *et al.* 1982; Lin and Savage 1984). Using two variants of *L. fermentum* 104, we have shown both non-specific and epithelial-specific adhesion to squamous epithelium *in vitro*. Proteins are involved in adhesion of both strains to epithelium as well as to polystyrene or BSA. The affinity for epithelium of 104-S is greater, compared to 104-R, however, 104-R has a higher affinity for the polystyrene (Fig. 5.2), (Henriksson *et al.* 1991). This reflects the difference in specificity of the mechanisms involved. Further studies of the 104-S show that the strain has a saccharide with a specific affinity for the epithelium (Henriksson and Conway 1992). In contrast, a protein from *L. fermentum* of rodent origin has been shown to adhere to rodent squamous tissue (Conway and Kjelleberg 1989). To date, the *in vivo* significance of these adhesion mediating components has not been investigated. Similarly, there are no published studies confirming the *in vivo* relevance of non-specific *in vitro* adhesion of lactobacillus.

Fuller (1973) presented excellent correlation between *in vitro* adhesive capacity and *in vivo* colonisation of lactobacilli in chicks. In contrast, Pedersen and Tannock (1989) found no correlation between the *in vitro* adhesive capacity of a series of lactobacillus strains and their potential for colonising the piglet gastrointestinal tract. Appropriate controls or a standardised washing procedure to remove non-specifically bound bacterial cells are essential. One can not conclude whether strains adhered specifically to the mucosa or whether the detected *in vitro* adhesion is non-specific and consequently may be less relevant *in vivo* unless these aspects are considered.

Using the well defined *E.coli* K12 ± K88 isogenic mutants, specific K88 fimbrial mediated adhesion and non-specific adhesion differed by a factor of 50-100 times (Table 5.2) in our system, while Cohen and co-workers (e.g. Metcalfe *et al.* 1991) frequently only obtain a 5-10 fold increased adhesion for *E. coli* K12+K88 compared to K12. Non-specific adhesion could conceivably lead to the attachment of significant numbers of lactobacilli to tissue *in vitro*. Numbers of adherent cells in the range of 5-50 fold greater than these controls might indicate the presence of specific adhesion. In attempting to address this issue, we have studied adhesion of an *L. fermentum* of human faecal origin to human, porcine, dog and monkey tissue gastrointestinal biopsy samples (Table 5.3). Although the strain adhered to porcine, dog and monkey tissue, enhanced adhesion to human tissue was noted with between 5 and 30 fold increase in numbers adhering to human tissue. This aspect was further investigated by studying the adhesion of human and porcine strains of *L. fermentum* to human gastric cells in tissue culture and porcine gastric epithelium (Table 5.4). The adhesion of the human strain to the human cells was 2-fold greater than that detected for the porcine strain 104 for the human cells. The porcine strain shows no indication of specific adhesion to the porcine tissue, relative to the human strain, and this is supported by published data showing that strain 104 exhibits a strong non-specific adhesion (Henriksson *et al.* 1991). The results in Tables 5.3 and 5.4 show a tendency for enhanced adhesion of the human strain to human tissue, suggestive of specific adhesion. One could extrapolate and propose therefore that this human strain may be equipped to colonise *in vivo*.

**Table 5.3.** Adhesion of *Lactobacillus fermentum* of human origin to gastrointestinal biopsy samples from a range of hosts

| | Log no of adhering bacteria[a] | | |
|---|---|---|---|
| Host | Tissue | KLD | No of samples |
| Monkey | Duodenum | 5.77 | 2 |
| | Jejunum | 5.33 | 2 |
| | Ileum | 5.35 | 2 |
| Dog | Stomach | 5.46 | 3 |
| Human | Stomach | 6.40 | 36 |
| | Duodenum | 6.79 | 21 |
| Pig | Stomach | 5.48 | 12 |

[a] Mean log of number of bacteria adhering per mg wet wt tissue (Standard deviation $\leq$ 12%)

**Table 5.4.** Adhesion to pieces of porcine gastric tissue and human AGS tissue culture cells of *L. fermentum* of human and porcine origin

| | Adhesion* | |
|---|---|---|
| Strain Origin | Human AGS-Cells (x $10^8$) | Porcine Tissue pieces (x $10^6$) |
| Human | 3.57 ± 0.23 | 3.85 ± 0.49 |
| Porcine | 1.42 ± 0.13 | 3.95 ± 0.43 |

* Number of attached cells per mg tissue (porcine) or per tissue culture well (human) ± SEM

### Antagonism Against Indigenous Microbiota

The potential of an administered microbe to colonise an ecosystem should also be discussed in terms of its interaction with the indigenous microbiota. Even if the introduced strain has the potential to specifically adhere to the epithelial mucosa, *in vivo* colonisation may not occur because it can not out-compete the already established microbes. Some workers studying colonisation, administer the probiotic strain after antibiotic treatment which suppresses the indigenous microbes (e.g. Lidbeck *et al.* 1988, Gotz *et al.* 1979). Alternatively, probiotic strains are selected which produce bacteriocins and related components which are inhibitory to the growth of related indigenous bacterial strains of the same genus (Klaenhammer 1988). This approach is based on the likelihood that related strains will occupy the same site in situ. For example, Harding and Shaw (1990) report antimicrobial activity of *Leuconostoc gelidum* against closely related species as well as against the pathogen *Listeria monocytogenes*.

### Persistence in situ

To develop successful strategies for functional probiotics, one must consider the environmental conditions to which the microbe is exposed after administration. This is particularly relevant for probiotics which must be viable and metabolically active in situ. The strain may be selected to survive in the target niche, for example the ileum, however, may not be equipped to survive transit through the hostile environment of the stomach and upper small intestine when administered by oral dosage. It is therefore important to consider the ecosystem in its entirety and select strains which can survive all the rigours such as low pH, bile acids and enzymes such as pepsin. A vast difference has been reported in the capacity of strains of lactic acid bacteria to survive low pH buffers and human gastric juice (Conway *et al.* 1987). It is therefore important to consider the resistance to in situ conditions of potential probiotic strains when discussing colonisation because without growth, colonisation can not be achieved, even for adhesive strains. The epithelial mucosa is continually regenerating and releasing older cells from the surface. Adhesive probiotic strains may well colonise a surface immediately after dosage but would be lost from the system via this sloughing process if the strain could not continue to grow and colonise the newly exposed epithelium.

## Biological Activity

To allow prediction that a probiotic preparation will be functional, it is obviously desirable to be able to demonstrate that it is biologically active against the target. Depending on the target, the activity of the probiotic can be discussed in terms of its antagonistic effects on microorganisms, or induced physiological and biochemical effects.

### Antagonistic Effects on Pathogenic Microorganisms

The inhibitory or antagonistic characteristics of probiotic strains are most often discussed in terms of their potential to produce metabolites that inhibit growth of pathogens as well as related strains which are constituents of the indigenous microbiota (eg Gilliland 1990). These metabolites can be low or high molecular weight compounds, the latter being referred to as bacteriocins when they are proteinaceous and plasmid encoded (Tagg *et al.* 1976). Numerous bacteriocins as well as low molecular weight inhibitory compounds produced by lactic acid bacteria have been reported (reviewed by Klaenhammer 1988; Fernandes *et al.* 1987). In addition to the organic acids produced by lactic acid bacteria, low molecular weight metabolites such as peptides produced by lactobacilli have been reported to be inhibitory to other bacterial strains (e.g. Silva *et al.* 1987). Similarly fatty acids metabolites from anaerobes such as *Bacteroides* can be inhibitory to enterotoxigenic *E. coli* (Wells *et al.* 1988). The spectrum of inhibitory activity varies considerably for the various reports. A metabolite with a very broad spectrum of antagonistic activity is the low molecular weight component reuterin produced by *Lactobacillus reuteri*. The purified reuterin inhibits growth of virtually all tested microorganisms (reviewed by Lindgren and Dobrogosz 1990). In contrast, very specific antagonistic effects have been reported. For example, the *Lactobacillus* sp tested by McCormick and Savage (1983) exhibited antagonistic activity towards *Clostridium ramosum*.

While growth inhibition has been extensively studied over the years, inhibition of the adhesion of the pathogen has been more recently addressed. This concept has been used for preventing pathogens adhering to a mucosal surface by pre-treating it with adhesive probiotic microorganisms. For example, Chan *et al.* (1985) prevented adhesion of uropathogenic *E. coli* to uro-epithelial cells by competitive exclusion by lactobacilli and cell wall fragments of lactobacilli. Similarly, Chauviere *et al.* (1992) described competitive exclusion by *L. acidophilus* cells of diarrhoea mediating *E. coli* to tissue culture cells. In addition, adhesion of pathogenic *E. coli* to intestinal cells was preventing by initially adhering lactobacilli to the ileal cells (Conway 1989). Recently, we extended the concept of pathogen inhibition by showing that high molecular weight components from growth of *Lactobacillus spp* of porcine origin can inhibit adhesion of pathogenic *E. coli* K88 (Conway *et al.* submitted). Furthermore, Kennedy and Volz (1985) inhibited gastrointestinal colonisation by *Candida albicans* through affecting adhesion, colonisation and dissemination using bacterial antagonism.

Inhibition of growth in situ or colonisation is also discussed in terms of competition for nutrients or a particular niche. For example, variants of *Clostridium difficile* which do not produce toxin have been shown to be effective in inhibiting toxigenic *C. difficile* (Seal *et al.* 1987). *C. difficile* is one of the most important causes of antibiotic associated diarrhoea and pseudo-membranous colitis. Gorbach and co-workers have successfully reduced relapses in patients with of *C. difficile* diarrhoea by dosing *L. acidophilus* GG (Goldin, personal communication).

## Physiological and Biochemical Effects

Reported physiological effects of probiotics include stimulation of the immune system, decreased incidence of tumours and improved gastrointestinal motility. There is increasing evidence that oral dosing of probiotic preparations of lactic acid bacteria can directly stimulate the immune system. Lactobacilli have been shown to function as an adjuvant (Gerritze *et al.* 1990), directly activating lymphocytes as well as liver and peritoneal macrophages and mononuclear phagocytes (Sato *et al.* 1988; Perdigon *et al.* 1986a,b). Gerritze *et al.* (1990) showed that oral dosage of trinitrophenylised (TNP) lactobacillus cells induced comparable to or even higher IgG anti-TNP serum titres than were obtained in intraperitoneally primed animals using a similar preparation of TNP cells. In contrast, decreasing the incidence of tumours and improving gastrointestinal motility are more indirect. A case study has been reported in which a constipated patient was treated with normal colon microbiota and the condition was reversed (Andrews *et al.* 1992). Evidence is accumulating that reductions in the incidence of tumours can be attributable to decreased activity of enzymes that can convert co-carcinogens to carcinogens e.g. azoreductase, nitroreductase (Goldin and Gorbach 1980, Goldin and Gorbach 1984). Another line of thought for the tumour suppressing effect of microorganisms is the induced host-mediated responses involving macrophage and lymphocyte activation (Kato *et al.* 1984) most probably activated by the immunopotentiating activities of lactic acid bacterial cell wall fractions (Bogdanov *et al.* 1975). In addition, probiotic strains can decrease levels of faecal mutagens (Lidbeck *et al.* 1991) most probably by binding the mutagen to the bacterial cell (Morotomi and Mutai 1986; Zhang and Ohta 1990). The biochemical effects that include deconjugation of bile acids as well as the assimilation of cholesterol and reduction of serum cholesterol are more tangible. The capacity of potential probiotic strains to exert these effects on bile acids and cholesterol can be relatively directly measured (Gilliland and Walker 1990).

## Evaluation of Probiotic Preparations

The natural progression for evaluating whether potential probiotic preparations function effectively is to initially use laboratory designed studies to establish if the strain can colonise and if it is biologically active against the target. By developing methodology to measure the biological activity, it may be possible to also obtain information about the mechanism of action. Provided the selected strain has no possibility of carrying virulence factors or other traits harmful to the host, the effectiveness of the preparations can then be evaluated in clinical studies. These can be used to initially confirm that the results of laboratory studies are also reproducible in the host (Stage 1) and subsequently evaluate the statistical effectiveness of the probiotic against the target on a larger scale (Stage 2)

### Laboratory Studies

#### *Colonisation*

As described earlier, colonisation can be discussed and therefore studied in terms of adhesion of the strain to the mucosa, antagonistic activity against the indigenous microbiota and the capacity of the strain to survive the rigours of the ecosystem. Evaluation of the adhesive nature of a strain has been discussed at length in the section in which the concept of colonisation is presented. To reiterate, an *in vitro* adhesion assay can be used provided the appropriate controls are included. The capacity of the strain to survive low stomach pH, bile acids and enzymes such as pepsins are relatively straight forward to measure *in vitro* (e.g. Conway *et al.* 1987). Antagonistic activities against growth of indigenous microorganisms can be studied as for antagonistic effects against growth of pathogens as described in the next section.

### *Biological Activity*

The production by the probiotic strain of metabolites which are inhibitory to the growth of other microbes has been extensively studied and appropriate methodology developed (e.g. Fernandez *et al.* 1978, Klaenhammer 1988, Juven *et al.* 1991). While these techniques can appear very convincing, it is vital to consider whether the antimicrobial activity demonstrated *in vitro* is stable and can function *in vivo*. We have detected considerable instability of the high molecular antagonistic metabolites in our digestive tract lactobacillus isolates and believe this is a relatively common phenomenon.

The biochemical effects including bile acid deconjugation and cholesterol assimilation can be tested *in vitro* and there seems to be an excellent correlation between *in vitro* assimilation of cholesterol and reduced serum cholesterol levels using the piglet as a model. The biological activity of the various physiological effects generally needs to be evaluated *in vivo* using a model system, e.g. mice or pigs.

## Clinical Studies

We have applied the strategies outlined here in the development of a probiotic preparation specifically designed for reducing piglet diarrhoea and mortality. Initially, we identified our target as the enterotoxigenic (ETEC) *Escherichia coli* bearing K88 fimbriae as a major cause of piglet pre- and post-weaning diarrhoea and death. A protocol was then developed for selecting strains with activity against this pathogen. In addition, the strains were screened for their capacity to adhere to epithelial mucosa, survive transit through, and to be metabolically active within the gastrointestinal tract. When one of the isolated strain was tested in a field trial, evidence of *in vivo* colonisation was demonstrated (Conway and Ronnow manuscript). Subsequently, in a field trial involving about 90 piglets per group, a significant reduction in diarrhoea levels and improved weight gain were detectable in the test group dosed with another isolate (Conway manuscript).

Similarly, we have isolated potentially functional probiotic strains from the faecal microbiota of healthy humans resistant to digestive tract disturbances when challenged by food-born pathogens. Strains were selected with demonstrable *in vitro* colonisation capacity and biological activity against a range of pathogens known to induce diarrhoea in humans (Conway unpublished data). Phase I clinical studies have been completed in human, non-patient volunteers and are described in Chapter 8, this volume.

## Summary

A considerable range of microorganisms are being investigated for therapeutic and prophylactic use in humans. Although earlier probiotic preparations were often largely ineffective, a new generation of probiotic preparations is emerging. The strategy of strain selection is established to ensure that the function and specificity of these probiotics is defined. Evidence is accumulating that infective conditions are most likely successfully treated by probiotics if the strain(s) are specifically selected in terms of colonisation potential and biological activity against the target. In addition, probiotic-induced biochemical effects hold promise because screening for appropriate strains and testing for biological activity may be feasible.

The treatment of physiological conditions such as tumours still presents a challenge largely because of the need to indirectly identify the target for the probiotic and therefore only study function by measuring related parameters. Recent studies of gastrointestinal disturbances yield evidence that several conditions previously believed to be physiologically induced, may in fact involve infective agents. This revelation allows consideration that such common gastrointestinal conditions as chronic constipation and diarrhoea, irritable bowel syndrome and inflammatory bowel disease may also be targeted for probiotic treatment. If infective agents are identified, it may be possible to select functional microorganisms with predictable and demonstrable probiotic action against these conditions.

In conclusion, by specifically defining the target for the probiotics, strict attention to correct methodology used for strain selection, and by ensuring *in vivo* biological activity, the cloud of uncertainty veiling probiotics should clear. Probiotics of the future need to be well characterised preparations with guaranteed effectiveness.

## *Acknowledgements*

The authors thank the Swedish Farmer's Supply and Crop Marketing Association for their financial support

## References

Alm L, Humble D, Ryd Kjellen E, Setterberg G (1983) The effect of acidophilus milk in geriatric patients. Nutrition and the intestinal flora, p 131 in Symp Swed Nutr Found 15 Bo Hallgren, ed, Uppsala

Andrews P, Barnes PJ, Borody TJ 1992. Chronic constipation reversed by reatoration of bowel flora. A case and a hypothesis Eur Journal of Gastroenterology and Hepatology, In Press

Asby CB, McEwan AD, Wilson SM, (1989) Effects of treatment of lambs with a probiotic containing lactic acid bacteria V Record 124:588-9

Barrow PA, Brooker BE, Fuller R, Newport MJ (1980) The attachment of bacteria to the gastric epithelium of the pig and its importance in the microecology of the intestine J Appl Bacteriol 48:147-154

Bhatia SJ, Kochar N, Abraham P, Nair NG, Metha AP (1989) *Lactobacillus acidophilus* inhibits growth of *Campylobacter pylori in vitro* J Clin Microbiol 27:2328-2330

Bhowmilk T, Johnson MC, Ray B (1985) Isolation and partial characterisation of the surface protein of *Lactobacillus acidophilus* strains. Int J Food Microbiol 2:311-321

Black FT, Andersen PL, Årskov J, Årskov F, Gaarslev K, Laulund S (1989) Prophylactic efficacy of lactobacilli on travellers diarrhoea Travel Medicine 5:333-335

Blomberg L, Conway PL (1989). An *in vitro* study of ileal colonisation resistance to *Escherichia coli* strain Bd 1107/75 08(K88) in relation to indigenous squamous gastric colonisation in piglets of varying ages Microbial Ecol in Health and Disease 2:285-291

Bogdanov IG, Dalev PG, Gurevich AI, Kolosov MN, Malókova VP, Plemyannikova LA, Sorokina IB (1975) Antitumor glycopeptides from *Lactobacillus bulgaricus* cell wall FEBS Lett 57:259

Borody TJ, George L, Andrews P, Brandi S, Noonan S, Cole P, Hyland L, Morgan A, Maysey J, Moore-Jones D (1989) Bowel-flora alternation: a potential cure for inflammatory bowel disease and irritable bowel syndrome? The Medical Journal of Australia 150:604

Chan RCY, Reid G, Irvin RT, Bruce AW, Costerton JW (1985) Competitive exclusion of uropathogens from human uroepitelial cells by Lactobacillus whole cells and cell wall fragments Infect Immun 47(1):84-89

Clements M L, Levine MM, Ristaino PA, Daya VE, Hughes TP (1983) Exogenous lactobacilli fed to man - their fate and ability to prevent diarrheal disease. Prog Fd Nutr Sci 7:29-37

Conway PL (1989) Lactobacilli: Fact and Fiction. In The regulatory and protective role of the normal microflora (ed Grubbe, Midvedt, Norin) MacMillan Press

Conway PL, Blomberg A, Welin A, Cohen P (1991). The role of piglet intestinal mucus and the pathogencity of *E. coli* K88. In: Molecular Pathogenesis of Gastrointestinal Infections (ed. Wadström et al.) Plenum pp 335-337

Conway PL, Gorbach SL, Goldin BR (1987) Survival of lactic acid bacteria in the human stomach and adhesion to intestinal cells J Dairy Sci 70:1-12

Conway PL, Kjelleberg S (1989) Protein mediated adhesion of *Lactobacillus fermentum* strain 737 to mouse stomach squamous epithelium J Gen Microbiol 135:1175-1186

Conway PL, Maki J, Mitchell R, Kjelleberg S (1986) Starvation of marine flounder, squid and laboratory mice and its effect on the intestinal microbiota FEMS Microbiol. Ecol 38:187-195

Conway PL, Welin A, Cohen SP (1990). The presence of K88 ab-specific receptor in porcine ileal mucus is age-dependent Infect Immun 58:3188-3182

Duval-Iflah Y, Ouriet MF, Moreau C, Daniel JC, Gabilan JC, Raibaud P (1982) Implantation precoce d'une souche de *Escherichia coli* dans l'intestin de nouveau-nes humains: effect de barriere vis-a-visde souches de E. coli antibioresistantes Ann Microbiol 133A:393-408

Fernandes CF, Shanhani KM, Amer MA (1987) Therapeutic role of dietary lactobacilli and lactobacillic fermented dairy products FEMS Microbiol Rev 46:343-356

Florey HW (1946) The use of microorganisms for therapeutic purposes Yale J Biol Med 19:101-117

Friend A, Farmer RE, Shahani KM (1982) Effect of feeding and intraperitoneal implatation of yoghurt culture cells on Ehrlich ascites tumor Milchwissenschaft 37(12):708-710

Fuller R (1973) Ecological studies on the lactobacillus flora associated with the crop epithelium of the fowl Journal of Appl Bacteriol 36:131-139

Fuller R (1992) Probiotics. Chapman and Hall

Gaastra W, de Graaf FK (1982) Host specific fimbrial adhesins of non-invasive enterotoxigenic *Escherichia coli* strains Microbiol Rev 46:129-161

Gerritse K, Posno M, Schellekens MM, Boersma WJA, Claassen E (1990) Oral administration of tnp-*Lactobacillus* conjugates in mice: A model for evaluation of mucosal and systemic immune responses and memory formation elicited by transformed lactobacilli Res Microbiol 141:955-962

Gilliland S E (1989) Acidophilus milk products: a review of potential benefits to consumers J Dairy Sci 72:2483-2494

Gilliland SE (1979) Beneficial interrelationships between certain microorganisms for use as dietary adjucts J Food Protection 42:164-167

Gilliland SE (1990) Health and nutritional benefits from lactic acid bacteria FEMS Microbiology Rev 87:175-188

Gilliland SE, Speck ML (1977) Deconjugation of bile acids by intestinal lactobacilli Appl Environm Microbiol 33(1):15-18

Gilliland SE, Walker DK (1990) Factors to consider when selecting a culture of *Lactobacillus acidophilus* as a dietary adjunct to produce a hypocholesterolemic effect in humans J Dairy Sci 73:905-911

Goldin BR (1986) In situ bacterial metabolism and colon mutagens Ann Rev Microbiol 40:367-393

Goldin BR, Gorbach SL (1980) Effect of *Lactobacillus acidophilus* dietary supplements on 1,2-dimethylhydrazine dihydrochloride induced intestinal cancer in rats J Natl Cancer Inst 64:263-265

Goldin BR, Gorbach SL (1984) The effect of milk and lactobacillus feeding on human intestinal bacterial enzyme activity American J. Clin Nutr 39:756-761

Goldin BR, Gorbach SL (1987) Lactobacillus GG: a new strain with properties favourable for survival, adhesion and antimicrobial activity in the gastrointestinal tract FEMS Microbiol Rev 46: p 72

Gorbach S L, Kean BH, Evans DG, Evans Jr. DJ, Bessudo D (1975) Travelers' diarrhea and toxigenic *Escherichia coli* New England Journal of Medicine 292:933-936

Gotz VP, Romankiewics JA, Moss J, Murray HW (1979) Prophylaxis against ampicillin-induced diarrhoea with a lactobacillus preparation Am J Hosp Pharm 36:754-757

Harding CD, Shaw BG (1990) Antimicrobial activity of *Leuconostaoc gelidum* against closely related species and *Listeria monocytogenes* J Appl Bacteriol 69:648-654

Henriksson A, Conway PL (1992) Adhesion to porcine squamous epithelium of saccharide and protein moieties of *Lactobacillus fermentum* strain 104-S. Journal of General Microbiology 138:2657-2661

Henriksson A, Sewzyk R, Conway PL (1991) Characteristics of the adhesive determinants of *Lactobacillus fermentum* 104 Appl Environ Microbiol 57(2):499-502

Hentges DJ (1983) Role of the intestinal microflora in host defense against infection. In: Human intestinal microflora in health and disease. (ed Hentges, D.J.) Academic Press, London, pp 311-331

Jonsson E, Conway PL (1992). Development of probiotics for pigs. In: Probiotics Ed. by R. Fuller, Chapman and Hall, pp 260-314

Juven BJ, Meinersmann RJ, Stern NJ (1991). Antagonistic effects of lactobaccilli and pediococci to control intestinal colonisation by human enteropathogens in live poultry. J Appl Bacteriol 70:95-103

Karlsson KA (1989) Animal glycosphingolipids as membrane attachment sites for bacteria Ann Rev Biochem 58:309-50

Kato I, Yokokura T, Mutai K (1983) Macrophage activation by Lactobacillus casei in mice. Microb Immunol 27:611-618

Kato I, Yokokura T, Mutai M (1984) Augmentation of mouse natural killer cell activity by Lactobacillus casei and its surface antigens. Microb Immunol 28:209-217

Kennedy MJ, Volz PA (1985) Ecology of *Candida albicans* gut colonisation: inhibition of candida adhesion, colonisation and dissemination from the gastrointestinal tract by bacterial antagonism Infect Immun 49(3):654-663

Klaenhammer TR (1988) Bacteriocins of lactic acid bacteria Biochemie 70:337-349

Klaenhammer TR, Kleeman EG (1981) Growth characteristics, bile sensitivity and freeze damage in colonial variants of *Lactobacillus acidopilus* Appl Environ Microbiol 41:1461-1467

Kleeman EG, Klaenhammer TR (1982) Adherence of lactobacillus species to human fetal intestinal cells J Diary Sci 65:2063-2069

Kurasz AB, Tanzer JM, Bazer L, Savoldi E (1986) In vitro studies of growth and competition between *S. salivarius* Tove-R and mutans streptococci J Dent Res 65:1149-1153

Lidbeck A, Allinger UG, Orrhage KM, Ottova L, Brismar B, Gustafsson J-Å, Rafter JJ, Nord CE (1991) Impact of *Lactobacillus acidophilus* supplements on the fecal microflora and soluble faecal bile acids in colon cancer patients Microbial Ecology in Health and Disease 4:81-88

Lidbeck A, Edlund C, Gustafsson JU, Kager L, Nord CE (1988) Impact of *Lactobacillus acidophilus* on the normal intestinal microflora after administration of two antimicrobial agents Infection 16(6):329-336

Lilly DM, Stillwell RN (1965) Growth promoting factor produced by microorganisms. Science 147:747-748

Lin JH-C, Savage DC (1984) Host specificity of the colonisation of murine gastric epithelium by lactobacilli FEMS Microbiol Letters 24:67-71

Lindgren SE, Dobrogosz WJ (1990) Antagonistic activities of lactic acid bacteria in food and feed fermentations FEMS Microbiol Rev 87:149-164

McCormick EL, Savage DC (1983) Characterisation of *Lactobacillus* sp. strain 100-37 from the murine gastrointestinal tract: ecology, plasmid content and antagonistic activity toward *Clostridum ramosum* H1 Appl Environ Microbiol 46:1103-1112

Metcalfe JW, Krogfelt KA, Krivan HC, Cohen PS, Laux DC (1991) Characterisation and identification of a porcine small intestine mucus receptor for the K88ab fimbrial adhesion Infect Immun 59:91-96

Metchnikoff E (1907) Quelques remarques sur le lait aigni. (Scientifically soured milk and its influence in arresting intestinal putrefication). GP Putman's Sons, New York

Mikkola R, Kurland CG (1991) Selection of laboratory wild type phenotypes from natural isolates of *E. coli* in chemostats Mol Biol Evol In press

Morotomi M, Mutai M (1986) In vitro binding of mutagenic pyrolyzates to intestinal bacteria Journal of the National Cancer Institute 77:195-201

Morotomi M, Watanabe T, Suegara N, Kawai Y, Mutai M (1975) Distribution of indigenous bacteria in the digestive tract of both conventional and gnotobiotic rats. Infect Immun 11:962-968

Muralidhara KS, Sheggeby GG, Elliker PR, England DC, Sandine WE (1977) Effect of feeding lactobacilli on the coliform and *Lactobacillus* flora of intestine tissue and feces from piglets J. Food Protect 40:288-296

Muralidhara KS, Sheggeby GG, Elliker PR, England DC, Sandine WE (1977) Effect of feeding lactobacilli on the coliform and lactobacillus flora of intestinal tissue and feces from piglets. J Food Protect 40:288-296

Oksanen PJ, Salminen S, Saxelin M, Hèmèlèinen P (1990) Prevention of travellers diarrhoea by *Lactobacillus* GG Ann Med 22:63-58

Pedersen K, Tannock GW (1989) Colonisation of the porcine gastrointestinal tract by lactobacilli Appl Environ Microbiol 55:279-283

Perdigon G, Alvarez S, Nader ME, de Macias , Margni RA, Oliver G, de R Holgado AP (1986a) Lactobacilli administrated orally induce release of enzymes from preitoneal macrophages in mice Milchwissenschaft 41:344-348

Perdigon G, Nader ME, Alvarez ME, Oliver G, Holgado AAPD, (1986b) Effect of perorally administered lactobacilli on macrophage activation in mice Infect Immun 53(2):404-410

Reid G, Cuperus PL, Bruce AW, van der Mei HC, Tomeczek L, Khoury AH, Busscher H (1992) Comparison of contact angles and adhesion to hexadecane of urogenital, dairy and poultry lactobacilli: Effect of serial culture passages Appl Environ Microbiol 58:1549-1553

Robins-Browne RM, Levine MM (1981) The fate of ingested lactobacilli in the proximal small intestine Am J Clin Nutr 34:514-519

Rolfe RD, Helebial S, Finegold SM (1981) Bacterial interference between *Clostridum difficile* and normal fecal flora J Infect Dis 143(3):470-475

Rutter PR, Dazzo FB, Freter R, Gingall D, et al (1984) Mechanisms of adhesion. In: Microbial Adhesion and Aggregation (Marshall, K.C. ed), Springer-Verlag, Berlin pp 5-19

Sato K, Saito H, Tomioka H (1988) Enhancement of host resistance against Listeria infection by *Lactobacillus casei*: activation of liver macrophages and peritoneal macrophages by *Lactobacillus casei* Microbiol Immunol 32(7):689-698

Savage DC (1972) Association and physiological interactions of indigenous microorganisms and gastrointestinal epithelia Am J of Clin Nutr 25:1372-1379

Saxelin M, Elo S, Salminen S, Vapaatalo H (1991) Dose response colonisation of faeces after oral administration of *Lactobacillus casei* strain GG Microbial Ecology in Health and Disease 4:209-214

Seal D, Borrielo SP, Barclay F, Welch A, Piper M, Bonnycastle M (1987) Treatment of relapsing *Clostridum difficile* diarrhoea by administration of a non toxigenic strain Eur J Clin Microbiol 6:51-53

Sears HJ, Janes H, Saloum R, Brownlee J, Lamoreau LF. (1956) Persistence of individual strains of *Escherichia coli* in man and dog under varying conditions J Bacteriol 71:370-372

Shinefield HR, Ribble JC, Boris M, Eichenwald HF (1963). Bacterial interference: its effect on nursery acquired infections with *Staphylococcus aureus* I, II, III, IV. Am J Dis Child 105:646-682

Silva M, Jacobus NY, Deneke C, Gorbach SL (1987) Antimicrobial substance from a human *Lactobacillus* strain. Antimicrobial Agents and Chemotherapy 31(8):1231-1233

Simon GL, Gorbach SL (1982) Intestinal flora in health and disease. In: Physiology of the gastrointestinal tract.(Johnson, L.R. ed), Raven Press New York, pp 1361-1380

Smith HW, Linggood MA (1971) Observations on the pathogenic properties of the K88, HLY and ENT plasmids of *Escherichia coli* with particular reference to porcine diarrhoea. J Med Microbiol 4:467-485

Tagg JR, Dajani AS, Wannamaker LW (1976) Bacteriocins of gram positive bacteria Bacteriol Rev 40:722-756

Tannock GW (1983). Effect of dietary and environmental stress on the gastrointestinal tract. Human intestinal microflora in health and disease. Hentges ed 1983 Academic Press, New York

Tannock GW (1990) The microecology of lactobacilli inhabiting the gastrointestinal tract Advances in Microbiol Ecology ed. K.C. Marshall 11:147-171

Tannock GW, Szylit O, Duval Y, Raibuad P (1982) Colonisation of tissue surfaces in the gastrointestinal tract of gnotobiotic animals by lactobacillus strains Can J Microbiol 28:1196-1198

Tvede M, Rask-Madsen J (1989) Bacteriotherapy for chronic relapsing *Clostridum difficile* diarrhoea in six patients The Lancet I:1156-1160

Wells CL, Maddaus MA, Jechorek RP, Simmons RL (1988) Role of intestinal anaerobic bacteria in colonisation resistance Eur J Clin Microbiol Infect Dis 7:107-113

Wilhelm MP, Lee DT, Rosenblatt JE (1987) Bacterial interference by anaerobic species isolated from human feces Eur J Clin Microbiol 6:266-270

Willcox MDP, Drucker DB (1988) Partial characterisation of the inhibitory substances produced by *Streptococcus oralis* and related species Microbios 55:135-145.

Winitz M, Adams RF, Seedman DA, Davis PN, Jakyo LG, Hamilton JA (1970) Studies in metabolic nutrition employing chemically defined diets. II Effects on gut microflora populations Am J Clin Nutr 32:546-559

Zhang XB, Ohta Y, Hosoni A (1990) Antimutagenicity and binding of lactic acid bacteria from a chinese cheese to mutagenic pyrolyzates J Dairy Sci 73:2702-2710

Ørskov I, Ørskov F, Smith HW, Sojka WJ (1975) The establishment of K99, a thermolabile transmissable Escherichia coli K antigen, previously called 'Kco', possessed by calf and lamb enteropathogenic organisms. Acta Path Microbiol. Scand 83:31-36

Chapter 6

# Lactobacilli, Normal Human Microflora and Antimicrobial Treatment

A. Lidbeck and C. E. Nord

## Normal Human Oropharyngeal and Gastrointestinal Microflora

### Development of the Oropharyngeal Microflora

Within the first few hours of life, microbial colonisation of the human body begins. In different ways, the microorganisms are transmitted via the mouth to sites of potential colonisation. This transmission originates from the mother during delivery, from feeding and from people coming into contact with the infant (Marsh and Martin 1984). These pioneer species consist mostly of aerobic and facultatively anaerobic bacteria. Already the day after birth, *Streptococcus salivarius* has been isolated from oral samples in most infants (Carlsson *et al.* 1975). Streptococci are the dominant oral microorganisms in newborn infants, together with staphylococci, neisseria, lactobacilli and veillonella (McCarthy *et al.* 1965). When the teeth are erupted, *Streptococcus sanguis* starts to colonise the mouth. Other microorganisms appearing at tooth eruption include *Streptococcus mutans* and *Actinomyces viscosus*. These microorganisms are mainly found in the dental plaque. At 1 year of age, *Actinomyces*, *Bacteroides* and *Fusobacterium* species have been isolated in about 50% of infants. This ecosystem containing only a few genera and species will further develop until a stable but dynamic system is reached (Marsh and Martin 1984; Macfarlane and Samaranayake 1989).

### Oropharyngeal Microflora in Healthy Adults

In saliva samples, microorganisms from different sites of the mouth are isolated, i.e. the cheek, tongue and teeth. In addition, microorganisms from the gingival crevices and tonsillar crypts are also found in mixed saliva (Mackowiak 1982; Heimdahl and Nord 1985). The number of bacteria in saliva is $10^8$ to $10^9$ colony-forming units

(CFU)/ml, and more than 300 different species have been isolated. The highest concentrations of anaerobic bacteria are found in habitats with a low oxidation-reduction potential, such as the dental plaque and the gingival crevice. In contrast, aerobic and facultatively anaerobic bacteria are found in high numbers on exposed surfaces of the teeth and on the epithelial tissues (Marsh and Martin 1984). The dominating oropharyngeal microorganisms are streptococci, peptostreptococci, lactobacilli, neisseria, veillonella, bacteroides and fusobacteria. *Candida albicans* can be isolated in 40% of healthy mouths but is usually found in low numbers.

## Development of the Gastrointestinal Microflora

In the gastric contents of 5-10 min old infants, the microflora is similar to the mother's cervical flora (Brook *et al.* 1979). However, some of the first microorganisms colonising the intestine after birth are *Escherichia coli* and streptococci. These aerobic or facultatively anaerobic organisms thereby create a reduced environment, favourable for the subsequent appearance of anaerobic microorganisms (Cooperstock and Zedd 1983). About 25 h after birth, the colon microflora in the newborn primarily consists of *Lactobacillus, Bifidobacterium* and *Bacteroides*. Clostridia also proliferate to moderately high levels during the first days of life. After the first one or two weeks of life, however, the number of clostridia rapidly declines to low or undetectable levels in breast-fed children (Bertazzoni *et al.* 1978; Nord and Kager 1984). At the same time, the initially high numbers of *E. coli* have been shown to decrease (Mata and Urrutia 1971; Patte *et al.* 1979). In a recent study by Hall *et al.* (1990), it was revealed that lactobacilli, but not bifidobacteria, were found in high counts in the stools of most infants by 30 days of age. In preterm infants a selective deficiency of lactobacilli compared to Gram-negative microorganisms was found.

The faecal microflora of children appears to resemble the adult flora by the age of one year. The numbers of aerobic microorganisms have decreased and anaerobes, particularly *Bacteroides* and *Lactobacillus*, become predominant in the faecal flora (Ellis-Pregler *et al.* 1975). At the age of 4-8 years, the microflora resembles that of adults. The same numbers of lactobacilli are seen as in adults and the ratio of anaerobic bacteria to aerobic bacteria is 1000:1.

## Impact of Feeding on the Microflora

Soon after birth breast-fed children obtain an intestinal microflora consisting mainly of bifidobacteria (Hentges 1980; Cooperstock and Zedd 1983).

Yoshioka *et al.* (1983) have studied the differences in the intestinal microflora in breast-fed and bottle-fed children. At one month of age, the breast-fed children harboured higher counts of bifidobacteria and lactobacilli and lower counts of Gram-negative bacteria such as enterobacteria and bacteroides compared to the bottle-fed infants.

In another study by Balmer and Wharton (1989), bifidobacteria and staphylococci were shown to dominate in breast-fed babies. The predominant organisms in the formula-fed infants were enterococci, clostridia and coliforms. In this study the protein content in a whey formula was 16% compared to only 9% in breast milk. The phosphorus and iron concentrations were also much higher in the formula milk. In a following study by Balmer *et al.* (1989), the effects on the faecal flora of whey proteins were compared to casein in formula-fed infants. When feeding a casein formula, more babies were colonised with *Bacteroides* at 14 days, while bifidobacteria dominated among the babies that were given whey protein and thus made the microflora resemble more that of breast-fed infants.

## Gastrointestinal Microflora in Healthy Adults

### *Upper Gastrointestinal Tract*

The microflora in the stomach resembles that of the oropharynx, since most microorganisms in the stomach are derived from the oral cavity and throat (Evaldson *et al.* 1982). Depending on the acidic environment in the stomach, the multiplication of most microorganisms is retarded, and mainly lactic acid bacteria are recovered. Gram-positive microorganisms such as streptococci, lactobacilli and bifidobacteria have been isolated from the stomach. Gram-negative bacteria represented by cocci, bacteroides and fusobacteria can also be found in the upper gastrointestinal tract. The numbers of bacteria in the stomach and the proximal small bowel are in the range of $10^2$-$10^5$ CFU/ml intestinal content. The number of bacteria in the stomach is regulated by the hydrochloric acid secretion. The number of microorganisms decreases when pH in the stomach falls. The number of bacteria may increase to $10^5$-$10^7$ CFU/ml after a meal, now also including bacteria from the food (Evaldson *et al.* 1982; Goldin 1986; Bergogne-Berezin 1989).

Small bowel motility prevents overgrowth of microorganisms. When the normal intestinal motility is inhibited, i.e. by obstruction or paralysis, there is a dramatic increase in the microbial population, and the microflora resembles that of the colon (Simon and Gorbach 1981). In the distal part of ileum, the microflora begins to resemble the colon microflora with numbers of $10^7$-$10^8$ bacteria/ml. *E. coli* dominates among the aerobic bacteria and is recovered in numbers of $10^7$-$10^8$ bacteria/ml.

**Table 6.1.** Composition of the human gastrointestinal microflora

| *Microorganisms* | *Number of microorganisms (CFU/ml or CFU/g)* | | | |
|---|---|---|---|---|
| | *Stomach* | *Jejunum* | *Ileum* | *Colon* |
| Total bacterial count | $0$-$10^3$ | $0$-$10^5$ | $10^3$-$10^9$ | $10^{10}$-$10^{12}$ |
| *Aerobic bacteria* | | | | |
| Enterobacteria | $0$-$10^2$ | $0$-$10^3$ | $10^2$-$10^7$ | $10^4$-$10^{10}$ |
| Streptococci | $0$-$10^3$ | $0$-$10^4$ | $10^2$-$10^6$ | $10^4$-$10^{10}$ |
| Staphylococci | $0$-$10^2$ | $0$-$10^3$ | $10^2$-$10^5$ | $10^4$-$10^9$ |
| Lactobacilli | $0$-$10^3$ | $0$-$10^4$ | $10^2$-$10^5$ | $10^6$-$10^{10}$ |
| Yeasts | $0$-$10^2$ | $0$-$10^2$ | $10^2$-$10^4$ | $10^4$-$10^6$ |
| *Anaerobic bacteria* | | | | |
| Bacteroides | rare | $0$-$10^3$ | $10^3$-$10^7$ | $10^{10}$-$10^{12}$ |
| Bifidobacteria | rare | $0$-$10^4$ | $10^3$-$10^9$ | $10^8$-$10^{11}$ |
| Streptococci | rare | $0$-$10^3$ | $10^2$-$10^6$ | $10^{10}$-$10^{12}$ |
| Clostridia | rare | rare | $10^2$-$10^4$ | $10^6$-$10^{11}$ |
| Eubacteria | rare | rare | rare | $10^9$-$10^{12}$ |

From Nord and Kager (1984)

### *Lower Gastrointestinal Tract*

In the colon the bacterial counts increase dramatically to numbers of $10^{10}$-$10^{12}$ CFU/g. Anaerobic bacteria, such as *Bacteroides* and *Bifidobacterium*, are the predominant species found in this region and the anaerobic bacteria outnumber the aerobic bacteria by a factor of 1000:1 (Nord and Kager 1984). One third of the faecal dry weight consists of bacteria. The composition of the gastrointestinal flora is presented in Table 6.1.

## Problems Associated with the Investigation of the Human Oropharyngeal and Gastrointestinal Microflora

When defining the oropharyngeal as well as the intestinal microflora, several problems have to be taken into consideration. As the composition of the flora is very complex and the microorganisms have fastidious growth requirements, it is of great importance that selective and differential media are used. A well functioning anaerobic technique is also essential, especially when looking for microorganisms present at low numbers.

When culturing saliva samples, it is important to standardise the sampling procedures according to the time of day, relation to food intake and tooth brushing, in order to get samples that are representative.

As it is difficult to obtain contents from the human colon, most studies have used faecal specimens. According to Moore *et al.* (1978), the colon contents are fairly well reflected by faecal specimens at least with respect to the microorganisms found there. Short chain fatty acids, as well as other compounds, are probably not the same as in the colon contents. The amount of water in the colon contents is at least twice as high as in the faeces. Adherent bacteria are more difficult to investigate from faecal samples, because only those which have been sloughed off will be represented in faeces. Tissue samples have indicated Gram-positive microorganisms as well as spirochetes attached to the colonic mucosa (Savage 1977; Finegold *et al.* 1983).

## Stability in the Ecology of the Intestinal Microflora

The normal human microflora is remarkably stable over time. Continually strong forces are exerted by the different populations to maintain the stability of the flora. Though certain variations exist in the composition of the microflora and in the numbers of microorganisms between individuals, the population sizes of the various bacteria from the same individual appear to be very stable (Simon and Gorbach 1981).

The number of microorganisms in the human upper gastrointestinal tract depends mainly on the rate of passage of nutrients and on the acidity of the stomach contents. The main characteristics of the normal gastrointestinal flora are well defined. Only certain particular organisms belong to the normal flora, and there is a geographic localisation of the specific organisms making the intestinal flora relatively constant over time (Bergogne-Berezin 1989). Certain factors that affect the colonic microflora are shown in Table 6.2.

Interactions occur between colonic mucosa and the bacterial population. Mucosal metabolic products with a beneficial effect on bacteria include oxygen, lactate and bicarbonate. Bacterial metabolic products which benefit mucosal cells include short chain fatty acids (acetate, propionate, butyrate), amines and other nitrogenous substances. The pH in the lumen depends both on the production of short chain fatty acids (SCFA) by microorganisms and the bicarbonate secretion which keeps the luminal pH at 7.4 with small variations (Roediger 1986).

**Table 6.2.** Factors affecting the colonic microflora

| | |
|---|---|
| Nutrient availability | Diarrhoea |
| Diet | Bacterial antagonism |
| pH | Bacterial co-operation |
| Redox potential | Mucin |

From Borriello (1986)

## Colonisation Resistance

Colonisation of the intestinal tract by potentially pathogenic microorganisms is usually prevented by interactions between the host and the normal microflora. Bacterial interference, both antagonistic and co-operative, probably plays a key role in the colonisation resistance against pathogenic microorganisms. Production of volatile fatty acids and bacteriocins as well as competition for attachment sites and nutrients contribute to the defence against pathogens. The intestinal anaerobic bacteria, and particularly the Gram-positive part of the microflora, are considered to be of importance for colonisation resistance (Savage 1977; van der Waaij 1983). Studies in humans by Gorbach *et al.* (1988) and Young (1988) have not supported this theory about colonisation resistance and its relationship to anaerobic microorganisms. However, Wells *et al.* (1988) reported that there was a significantly increased rate of translocation of aerobic bacteria to the mesenteric lymph nodes in mice when anaerobic bacteria were absent. There is also still a lack of information concerning the role of indigenous lactobacilli in the control of other intestinal microorganisms. The mechanisms, by which certain populations can affect the population sizes of other species, have been studied by Itoh and Freter (1989). In an animal study with gnotobiotic mice, lactobacilli suppressed *E. coli* multiplication in the stomach and small intestine, but appeared to have no effect on the number of *E. coli* in the colon. The control of *E. coli* populations was not related to changes in pH or intestinal motility. In the colon, clostridia were most effective in controlling *E. coli* multiplication.

## Role of Normal Microflora in Disease

Opportunistic infections may be caused by many species in the normal flora when they invade protected areas of the body or if the host defence mechanisms are compromised. In the oral cavity, members of the normal flora cause dental caries and periodontal diseases. At tooth extraction, viridans streptococci from the oral cavity invade the bloodstream and may cause endocarditis if the heart valves are damaged.

If the immune response is reduced or the epithelial cell barriers are weakened by nutrient deficiencies, the normal flora may invade the tissues and cause infections.

Immunocompromised patients and cancer patients treated with chemotherapy often develop infections caused by microorganisms belonging to the normal microflora.

### Impact of Antimicrobial Therapy

Therapy with antimicrobial agents may cause pronounced disturbances in the normal microflora (Nord *et al.* 1986). Antimicrobial agents are important in the treatment and prophylaxis of infections. However, it should be considered that some of these agents have a harmful effect on the human microflora, leading to undesired effects such as overgrowth and superinfections with commensal microorganisms, such as yeasts. A great deal of surgical infections appearing during antimicrobial treatment are caused by Gram-negative aerobic and anaerobic rods belonging to the normal intestinal microflora. There has been a change in responsible microorganisms in surgical infections from invading exogenous pathogens to potentially pathogenic indigenous microorganisms (Nord 1990). A recent report concerning the relative occurrence of pathogens in 900 intraabdominal infections revealed that *E. coli, Bacteroides fragilis* and *Clostridium* spp. accounted for the major part of the causative bacterial species (Wittman 1991).

Incomplete absorption of orally administered drugs may lead to a suppression of susceptible microorganisms and thus disturb the ecological balance in the intestine. Parenteral administration of agents secreted by the salivary glands, in the bile or from the intestinal mucosa can also interrupt the normal microbial population (Nord *et al.* 1986). The potential of an antimicrobial agent to change the normal microflora is dependent on its dose, pharmacokinetic properties and route of administration.

Suppression of the normal microflora lowers the colonisation resistance, often leading to establishment of potentially pathogenic microorganisms such as *Candida* species. These pathogens are often resistant to the antimicrobial agent used and may cause stomatitis, diarrhoea or colitis. Furthermore, in immunocompromised patients *Candida* may cause systemic infections. Overgrowth by toxin-producing *Clostridium difficile* can give rise to diarrhoea, colitis and pseudomembranous colitis. Antimicrobial agents such as cephalosporins, clindamycin and ampicillin have been associated with this disease (Aronsson *et al.* 1985).

## Human Lactobacillus Microflora

### Lactobacilli in Human Ecology

Colonisation of the human body with different microorganisms begins at delivery. Lactobacilli constitute the dominant part of the vaginal microflora and thus provide an inoculum to the infant during birth (Bartlett *et al.* 1977). The same kind of lactobacilli, i.e. *L. acidophilus* and *Lactobacillus jensenii* have been recovered in vaginal samples from the mother and oral samples from the infant (Carlsson and Gothefors 1975), although there is no evidence that these lactobacilli colonise the infant digestive tract (Tannock *et al.* 1990). The lactobacilli and streptococci, which are established in the mouth of infants, follow a specific pattern. Those lactobacilli acquired at birth gradually disappear and other *Lactobacillus* sp. are detected (McCarthy *et al. 19*65). About 25 h after birth, the colon microflora in the newborn primarily consists of lactobacilli and bifidobacteria. The following lactobacilli are usually recovered in infant stools: *L. acidophilus, L. salivarius,* and *L. fermentum.* The same *Lactobacillus* sp. are also found in adults, together with additional lactobacilli transiently harboured in the intestine (Cooperstock and Zedd 1983).

In elderly persons the numbers of lactobacilli and clostridia are significantly higher and the number of bifidobacteria lower than those seen in younger adults (Speck 1976). In the Wadsworth study (Finegold *et al.* 1983), *Lactobacillus* sp. were found in 73% of individuals eating a "Western" diet, with a mean count of Log 9.3 organisms per gram dry weight faeces (range Log 3.6 - Log 12.5). In strict vegetarians, lactobacilli were recovered in 85% with a mean value of Log 11.1 (range Log 8.6 - Log 12.1). *L. acidophilus* was the most frequently isolated species and was recovered in 44.7% of these subjects.

## Lactobacilli in Oropharyngeal Ecology

Lactobacilli are commonly found in low numbers in dental plaque, usually comprising less than 1% of the total cultivable flora. Several species of lactobacilli have been isolated from the mouth. The most frequently found species of lactobacilli appear to be *L. casei* and *L. fermentum* (Macfarlane and Samaranayake 1989). These microorganisms are very acidogenic, and earlier lactobacilli were believed to be the causative agent of dental caries (Table 6.3).

**Table 6.3.** Factors related to the cariogenicity of *Lactobacillus* species

| |
|---|
| High numbers in most carious lesions affecting enamel |
| Numbers in plaque and saliva often positively correlated with caries activity |
| Some strains produce caries in gnotobiotic rats |
| Able to initiate and maintain growth at low pH levels |
| Produce lactic acid in conditions below pH 5.0 |
| Some strains synthesise both extra- and intracellular polysaccharides from sucrose |

From Macfarlane and Samaranayake (1989)

Children with active dental caries often had higher numbers of lactobacilli in saliva compared to caries-free children. However, more carefully controlled studies have shown that some subjects with high caries activity have low numbers of salivary lactobacilli, while caries-free individuals sometimes have high numbers of lactobacilli in the saliva. Later studies indicated that *Streptococcus mutans* was more correlated to the development of dental caries. *S. mutans* is the most acidogenic of the oral streptococci and produces acid faster and in greater quantities than lactobacilli. The resulting lower pH makes the environment favourable for lactobacilli, which may colonise and produce some acid, thereby being involved in the dental caries process. Their affinity for the tooth surface and their numbers in dental plaque associated with healthy sites are usually low. In addition, lactobacilli are rarely isolated from plaque prior to the development of caries .

Lactobacilli are usually considered as non-pathogenic microorganisms. However, according to clinical reports, *L. casei* and *L. plantarum* have been involved in human infections. *Lactobacillus* spp. have been reported to cause infective endocarditis of oral origin (Sussman *et al.* 1986).

## Lactobacilli in Intestinal Ecology

The microflora in the oropharynx and the stomach is similar, since most microorganisms in the stomach are derived from the oral cavity and throat. Lactobacilli in the stomach are found in numbers of $10^3$-$10^4$ CFU/g together with streptococci and bifidobacteria. Depending on the acidic environment in the stomach, the multiplication of most microorganisms is retarded and mainly lactic acid bacteria are recovered. Small bowel motility also prevents overgrowth of microorganisms, but in the distal part of the ileum there is a significant increase in bacterial counts and lactobacilli are recovered in numbers of $10^3$-$10^7$ CFU/g. *E. coli* dominates among the aerobic microorganisms and is isolated in numbers of $10^7$-$10^8$ CFU/g. In the colon the bacterial counts increase further to $10^{10}$-$10^{12}$ CFU/g and the anaerobic bacteria outnumber the aerobic bacteria by a factor of 1000:1. Lactobacilli are recovered in numbers of $10^4$-$10^8$ CFU/g, while *B. fragilis* is the dominating anaerobic microorganism (Evaldson *et al.* 1982; Nord and Kager 1984; Goldin 1986).

## Adherence and Survival of Lactobacilli in the Gastrointestinal Tract

Animal studies have shown that the capacity of the microorganisms to colonise the epithelial surface may be specified as much by a capacity to bind to the epithelium as by nutritional and environmental conditions. Animal studies have indicated that lactobacilli adhere to squamous epithelia via acidic polysaccharides. However, it has been reported that macromolecules other than polysaccharides may be involved in the adhesion (Savage 1984).

Indigenous lactobacilli are believed to attach to the epithelium of the intestine by an adherence mechanism (Savage 1979). Kleeman and Klaenhammer (1982) tested a number of lactobacilli for their ability to adhere to a human fetal intestinal epithelial cell line. Two mechanisms of adhesion were found; one was calcium requiring and non-specific allowing all the tested lactobacilli to adhere. In this system, *L. acidophilus* NCFM - a commercial strain - showed a strong adhesion capacity. The other, non-calcium requiring mechanism was found in 4 human *L. acidophilus* isolates. This adherence was not affected by broth passage, freezing or lyophilisation.

In a study by Conway *et al.* (1987), the ability of different strains of lactobacilli to survive in gastric juice was tested in healthy subjects. (Three subjects received *L. acidophilus* strain N 2 and 2 subjects *L. acidophilus* strain ADH and *L. bulgaricus*, respectively.) The survival of the lactobacilli was related to pH, and the strain showing the best survival in gastric juice was *L. acidophilus* strain ADH. This strain also had the best adhesion capacity to human ileal cells. The addition of milk to the supplement caused a rise in pH and increased survival times for all lactobacilli in gastric juice.

Robins-Browne and Levine (1981) have shown that *L. acidophilus* and *L. bulgaricus*, when taken with milk, had the ability to pass through the stomach in equal counts. Jejunal fluid samples were taken from volunteers at varying intervals, and lactobacilli were cultured on a selective medium. When the volunteers were fasting, the lactobacilli survived for about 3 h in the small intestine. In non-fasting subjects they survived for almost 6 h and *L. acidophilus* was recovered from more samples and in somewhat higher counts than *L. bulgaricus*. The survival rate was found to be 1.3%

when *L. acidophilus* NCFB (formerly NCDO) 1748 was tested for the ability to pass the stomach and intestine *in vivo* (Pettersson *et al.* 1983).

## Disturbances of the Protective Microflora

The normal gastrointestinal microflora is relatively constant during lifetime, but certain factors may affect this equilibrium. Dietary and environmental conditions can influence this ecosystem (Tannock 1983). As a result of emotional stress in humans, the secretion of hydrochloric acid in the stomach may decrease. Normally low numbers of bacteria, mostly lactobacilli and streptococci are found in the stomach, but subjects with gastric achlorhydria may harbour high numbers of coliforms and *Bacteroides* (Drasar *et al.* 1969). In children suffering from protein-calorie malnutrition, lactobacilli are found in lower counts and coliforms in higher counts compared to normal subjects (Tannock 1983). Thus the number of lactobacilli could apparently be affected by dietary stress.

**Table 6.4.** Effect of antimicrobial agents on *Lactobacillus* counts in the intestinal microflora

| Agent | No of pat* | Dose | Days of treatment | *Lactobacillus* | Reference |
|---|---|---|---|---|---|
| Phenoxymethyl-penicillin | 3 | 1-2 g/day p.o. | 7 | Elim** 2 pat | Finegold *et al.* (1967) |
| Ampicillin | 16 | 1.5 g/day p.o. | 5 | Decrease 2-3 logs 16 pat Elim 9 pat | Daikos *et al.* (1968) |
| Ampicillin + Sulbactam | 20 | 1.5 g/day i.v. | 2 | Reduction in count | Kager *et al.* (1982) |
| Cephalexin | 3 | 2 g/day p.o. | 6-7 | Reduction in count | Sutter and Finegold (1974) |
| Moxalactam | 5 | 0.5 g q8h i.v. | 10 | Decrease ≥4 logs 2 pat | Allen *et al.* (1980) |
| | 5 | 2 g q8h i.v. | 10 | Decrease ≥4 logs 3 pat | |
| Cefmenoxime | 15 | 4 g/day i.v. | 3 | Decrease 3-6 logs | Knothe *et al.* (1985) |
| Cefoperazone | 4 | 4 g/day i.v. | 12-30 | Elim 3 pat | Mulligan *et al.* (1982) |
| | 29 | 2 g x 2 i.v. | 7-14 | Decrease | Alestig *et al.* (1983) |
| Clindamycin | 5 | 1.2 g/day p.o. | 6-12 | Elim 2 pat | Sutter and Finegold (1974) |
| | 6 | 500mg q2h x2 then 250mg q6h | 3 | Elim | Bornside and Cohn (1969) |
| | 10 | 150 mg x4 = 600mg/ day | 7 | Decrease 5 pat Elim 2 pat | Lidbeck *et al.* (1988) |
| Chloramphenicol | 25 | 1.5 g/day p.o. | 5 | Decrease | Daikos *et al.* (1968) |
| Neomycin | 10 | 2 g/day p.o. | 6 | Decrease | Daikos *et al.* (1968) |
| | 4 | 4 g/day p.o. | 7 | 1-2 logs | |
| Neomycin + Metronidazole | 9 | 1 g neomycin + 200 mg metronidazole q8h | 5 | Decrease almost 5 logs | Arabi *et al.* (1979) |

Pat* = patients; elim** = elimination p.o. = per oral administration; i.v. = intravenous administration

## Impact of Antimicrobial Agents on Lactobacilli

In many cases antimicrobial therapy is accompanied by gastrointestinal disturbances and either a reduction or an elimination of lactobacilli in the intestinal microflora (Heimdahl and Nord 1979; Finegold *et al.* 1983). Cephalosporins given orally or intravenously decreased the number of lactobacilli by more than 2 logs (Allen *et al.* 1980; Mulligan *et al.* 1982; Alestig *et al.* 1983; Ambrose *et al.* 1985; Knothe *et al.* 1985). In several studies clindamycin has been shown to eliminate or reduce the number of lactobacilli (Sutter and Finegold 1974; Allen *et al.* 1980; Lidbeck *et al.* 1988). Chloramphenicol reduced the lactobacilli counts by 2 and 3 log cycles (Daikos *et al.* 1968). The administration of neomycin or a combination of neomycin and metronidazole also decreased the number of lactobacilli significantly (Daikos *et al.* 1968; Arabi *et al.* 1979) (Table 6.4).

## Attempts to Re-establish the Microflora with Lactobacilli

In a number of previous studies lactobacilli have been used in attempts to prevent diarrhoea associated with antimicrobial agents (Beck and Necheles 1961; Pearce and Hamilton 1974). In order to prevent ampicillin-associated diarrhoea, Gotz *et al.* (1979) gave a mixture of *L. acidophilus* and *L. bulgaricus* to one group of patients, while the other group received placebo. When the patients with diarrhoea unrelated to ampicillin were excluded, the incidence of ampicillin-related diarrhoea in the placebo group was significantly higher compared to the *Lactobacillus* group.

The effect of the same *Lactobacillus* mixture was studied for its possible efficacy in preventing neomycin-associated diarrhoea. One batch of the preparation reduced the frequency and severity of diarrhoea in volunteers, while no protective effect could be detected with the other batch (Clements *et al.* 1983). In a recent study by Tankanow *et al.* (1990), the same freeze-dried preparation of *L. acidophilus* and *L. bulgaricus* did not appear to consistently prevent amoxicillin-induced diarrhoea in children. Thus, the results obtained with this kind of *Lactobacillus*-preparation have been contradictory.

Different lactobacillus preparations were given to infants treated with ampicillin (Zoppi *et al.* 1982). A normalisation of the equilibrium of the intestinal bacteria was obtained when *L. acidophilus* plus *Bifidobacterium bifidum* were given. The number of lactobacilli increased significantly and a reestablishing effect of these microorganisms was reported. When another preparation containing *L. acidophilus* plus *Streptococcus lactis* was given, a partial restoration of the microflora occurred with no increase in the number of lactobacilli.

Yost and Gotz (1985) evaluated the effect of a combination of *L. acidophilus* and *L. bulgaricus* on the absorption of ampicillin. No differences were seen in maximum plasma concentrations of ampicillin or in half lives, whether the lactobacilli were given concurrently or not. Thus, the lactobacilli did not affect the absorption of orally given ampicillin.

The impact of *L. acidophilus* on the oropharyngeal and intestinal microflora has been studied before, during and after *L. acidophilus* fermented milk administration (Lidbeck *et al.* 1987). It was given in a dose of 250 ml twice daily for one week to ten healthy volunteers. Each ml of the product contained $5x10^8$-$2x10^9$ CFU of *L. acidophilus* NCFB (formerly NCDO) 1748. In the oropharynx only minor changes in the microflora were observed and there were no increase in lactobacilli. In the intestinal microflora there was an increase in the number of lactobacilli in nine of ten subjects while a decrease in the number of *E. coli* was observed in six subjects. The increase in lactobacilli remained as long as the subjects were consuming the *L. acidophilus* supplementation and should thus be taken continuously to keep the higher levels of lactobacilli in the intestine.

When clindamycin was given to ten healthy volunteers (Lidbeck *et al.* 1988) the anaerobic microflora was strongly suppressed. In two subjects lactobacilli disappeared and a decrease occurred in five subjects. *L. acidophilus* supplementation for seven days to five subjects resulted in a significant increase in numbers of lactobacilli in all those subjects. Most other anaerobic bacteria returned to the same levels as before clindamycin administration one week later. *Candida albicans* was detected in eight subjects, four in each group, during clindamycin administration. During *L. acidophilus* supplementation *C. albicans* disappeared in three of four subjects in the *Lactobacillus* group, while no similar decrease occurred in the other group. These findings indicate that *L. acidophilus* administration might lower the risk of *Candida* infections in compromised patients.

In premature infants, the effect of *L. acidophilus* feeding on colonisation by antibiotic-resistant Gram-negative microorganisms was determined. One group of infants received a lactobacilli-containing formula and the other group a formula with no lactobacilli. There was no significant difference between the two groups concerning the number of aminoglycoside resistant bacteria (Reuman *et al.* 1986).

Lactobacilli have also been tested for the prevention of traveller's diarrhoea (Clements *et al.* 1981). The volunteers were challenged with enterotoxigenic *E. coli* and given either a *Lactobacillus* preparation or placebo. There was no significant difference in clinical symptoms between the 2 groups. Recently Black *et al.* (1989) showed that a mixture of 4 different bacterial species, *L. acidophilus* plus *B. bifidum* (90%) *L. bulgaricus*, and *Streptococcus thermophilus*, significantly reduced the frequency of diarrhoea from 71% to 43% in tourists visiting Egypt.

The same preparation of lactic acid producing bacteria was used in a recent double-blind study concerning the effect of lactic acid producing bacteria on the intestinal microflora during ampicillin administration (Black *et al.* 1991). The volunteers receiving the mixture of lactic acid bacteria were recolonised faster with *E. coli* compared to the placebo group. Higher counts of anaerobic Gram-positive cocci, bifidobacteria, eubacteria and lactobacilli were found in the subjects receiving the lactic acid producing bacteria. Adverse effects such as diarrhoea were observed in three subjects in the ampicillin-placebo group compared to one subject in the lactic acid bacteria group.

In a double-blind study *Lactobacillus* GG was given in a daily dose of $2 \times 10^9$ CFU in order to investigate a possible effect of lactobacilli against traveller's diarrhoea during a trip to southern Turkey. In the placebo group the total incidence of diarrhoea was 46.5% compared with 41% in the *Lactobacillus* GG group (Oksanen *et al.* 1990).

## Lactobacillus acidophilus and Fermented Milk Products

Fermentation with lactic acid bacteria is a very old method of food preservation and lactobacilli have been used together with streptococci in the manufacture of dairy products. Since enzymes of lactic acid cultures degrade proteins, lipids and lactose of milk, these nutritional components are partially predigested (Alm 1982). Almost every civilisation has consumed cultured milks and these products have been, and still are, of great importance in the nutrition of people throughout the world (Shahani 1983). Many lactic cultures require vitamins for growth, but they are on the other hand also capable of synthesising B-vitamins during fermentation. Particularly folic acid is found in higher levels in cultured dairy products.

The establishment of certain lactobacilli in the gastrointestinal tract is believed to contribute to the stabilisation of the microflora of healthy subjects and to exert beneficial effects on the host (Metchnikoff 1908; Smith 1924; Rettger *et al.* 1935; Speck 1976; Sandine 1979; Fernandes *et al.* 1987). Because of these proposed healthful properties, several *Lactobacillus* sp., especially *L. acidophilus*, have been used both as dietary supplements and in attempts to prevent gastrointestinal disturbances.

Lactobacilli belong to the Gram-positive non-sporing facultative or anaerobic rods. These microorganisms utilise carbohydrates as main nutritional source and are found in fermenting animal and plant products. The main end product of glucose fermentation is lactic acid, resulting in a decrease of pH, by one or more units. Besides lactic acid, lactobacilli also produce acetic acid and hydrogen peroxide. These metabolites make the environment less favourable for the *in vitro* growth of potentially pathogenic microorganisms, such as staphylococci, pseudomonas and salmonella (Mehta *et al.* 1983; Speck 1983). Recently Bhatia *et al.* (1989) have shown that *L. acidophilus* inhibited the growth of seven isolates of *Helicobacter pylori in vitro*, probably due to the production of lactic acid.

*L. acidophilus, L. bulgaricus, L. casei* and *L. plantarum* belong to the homofermentative lactobacilli, producing more than 85% lactic acid from glucose. The heterofermentative lactobacilli produce at least 50% lactic acid together with acetic acid, ethanol and carbon dioxide. *Lactobacillus brevis, Lactobacillus cellobiosus* and *L. fermentum* are found in this group (Kandler and Weiss 1986). Several other substances with antimicrobial properties are also produced (Shahani *et al.* 1977; Barefoot and Klaenhammer 1983; Silva *et al.* 1987; Axelsson *et al.* 1989).

The nutritional requirements for *L. acidophilus* are complex, including acetate or mevalonic acid, riboflavine, pyridoxine, carbohydrates and amino acids (Kandler and Weiss 1986). *L. acidophilus* produces DL-lactic acid from lactose, glucose, maltose, saccharose and other carbohydrates. The colonies have an irregular shape and are uncoloured. The optimum temperature for growth is 35°C - 38°C, and the optimum pH for growth is pH 5.5-6.0.

Important selection criteria of strains used in the production of fermented acidophilus milk products should be safety, technological properties as well as beneficial effects concerning the influence on the intestinal flora (Fondén 1989). The strain selected for a particular purpose should be chosen on the basis of its ability to produce the desired effect at a maximum level (Gilliland 1989). In addition to possessing the ability to produce the desired effect, it is important that the selected microorganism possesses certain other desirable characteristics. One important factor is the ability to survive, grow and perhaps establish itself in the gastrointestinal tract. It is also important that it has the ability to compete with other lactobacilli naturally occurring in the intestine. The production of bacteriocins by the lactobacilli may thus also be of importance for the growth and establishment in the intestinal tract.

Another important factor for the survival and growth of the selected lactobacilli is bile resistance. These characteristics must be maintained by the selected organism during production and storage of the cultured milk product.

# References

Alestig K, Carlberg H, Nord CE, Trollfors B (1983) Effect of cefoperazone on faecal flora. J Antimicrob Chemother 12:163-167

Allen SD, Siders JA, Cromer MD, Fischer JA, Smith JW, Israel KS (1980) Effect of LY 127935 (6059-S) on human fecal flora. In: Nelson JD, Grassi C, eds. Current chemotherapy and infectious disease. Vol. 1. Washington: The American Society for Microbiology pp 101-103

Alm L (1982) The effect of fermentation on nutrients in milk and some properties of fermented liquid milk products. Thesis, Karolinska Institute, Stockholm

Ambrose NS, Johnson M, Burdon BW, Keighley MR (1985) The influence of single dose intravenous antibiotics on faecal flora and emergence of *Clostridium difficile*. J Antimicrob Chemother 15:319-326

Arabi Y, Dimock F, Burdon DW, Alexander-Williams J, Keighley MRB (1979) Influence of neomycin and metronidazole on colonic microflora of volunteers. J Antimicrob Chemother 5:531-537

Aronsson B, Möllby R, Nord CE (1985) Antimicrobial agents and *Clostridium difficile* in acute enteric disease: Epidemiological data from Sweden 1980-1982. J Infect Dis 151:476-481

Axelsson LT, Chung TC, Dobrogosz WJ, Lindgren SE (1989) Production of a broad spectrum antimicrobial substance by *Lactobacillus reuteri*. Microb Ecol Health Dis 2:131-136

Balmer SE, Scott PH, Wharton BA (1989) Diet and faecal flora in the newborn: casein and whey proteins. Arch Dis Child 64:1678-1684

Balmer SE, Wharton BA (1989) Diet and faecal flora in the newborn: breast milk and infant formula. Arch Dis Child 64:1672-1677

Barefoot SF, Klaenhammer TR (1983) Detection and activity of Lactacin B, a bacteriocin produced by *Lactobacillus acidophilus*. Appl Environ Microbiol 45:1808-1815

Bartlett JG, Onderdonk AB, Drude E, Goldstein C, Anderka M, Alpert S, McCormack WM (1977) Quantitative bacteriology of the vaginal flora. J Infect Dis 136:271-277

Beck C, Necheles H (1961) Beneficial effects of administration of *Lactobacillus acidophilus* in diarrheal and other intestinal disorders. Am J Gastroenterol 35:522-530

Bergogne-Berezin E (1989) Microbial ecology and intestinal infections. In: Bergogne-Berezin E, ed. Microbial ecology and intestinal infections. Paris: Springer-Verlag pp 1-5

Bertazzoni EM, Benoni G, Berti T, Deganello A, Zoppi G, Gaburro D (1978) A simplified method for the evaluation of human faecal flora in clinical practice. Helv Paediat Acta 32:471-478

Bhatia SJ, Kochar N, Abraham P, Nair NG, Mehta AP (1989) *Lactobacillus acidophilus* inhibits growth of *Campylobacter pylori* in vitro. J Clin Microbiol 27:2328-2330

Black FT, Andersen PL, Ørskov J, Ørskov F, Gaarslev K, Laulund S (1989) Prophylactic efficacy of lactobacilli on traveler's diarrhea. In: Steffen R, ed. Travel medicine. Conference on international travel medicine 1, Zürich, Switzerland, 1988. Berlin: Springer-Verlag pp 333-335

Black F, Einarsson K, Lidbeck A, Orrhage K, Nord CE (1991) Effect of lactic acid producing bacteria on the human intestinal microflora during ampicillin treatment. Scand J Infect Dis 23:247-254

Bornside GH, Cohn I Jr (1969) Intestinal antisepsis: Stability of fecal flora during mechanical cleansing. Gastroenterology 57:569-573

Borriello SP (1986) Microbial flora of the gastrointestinal tract. In: Hill MJ, ed. Microbial metabolism in the digestive tract. Florida: CRC Press pp 1-19

Brook I, Barett C, Brinkman C, Martin W, Finegold SM (1979) Aerobic and anaerobic bacterial flora of the maternal cervix and newborn gastric fluid and conjunctiva: A prospective study. Pediatrics 63:451-455

Carlsson J, Gothefors L (1975) Transmission of *Lactobacillus jensenii* and *Lactobacillus acidophilus* from mother to child at time of delivery. J Clin Microbiol 1:124-128

Carlsson J, Grahnén H, Jonsson G (1975) Lactobacilli and streptococci in the mouth of children. Caries Res 9:333-339

Clements ML, Levine MM, Black RE, Robins-Browne RM, Cisneros LA, Drusano GL, Lanata CF, Saah AJ (1981) Lactobacillus prophylaxis for diarrhea due to enterotoxigenic *Escherichia coli*. Antimicrob Agents Chemother 20:104-108

Clements ML, Levine MM, Ristaino PA, Daya VE, Hughes TP (1983) Exogenous lactobacilli fed to man - their fate and ability to prevent diarrheal disease. Prog Food Nutr Sci 7:29-37

Conway PL, Gorbach SL, Goldin BR (1987) Survival of lactic acid bacteria in the human stomach and adhesion to intestinal cells. J Dairy Sci 70:1-12

Cooperstock MS, Zedd AJ (1983) Intestinal flora of infants. In: Hentges DJ, ed. Human intestinal microflora in health and disease. New York: Academic Press pp 79-99

Daikos GK, Kontomichalou P, Bilalis D, Pimenidou L (1968) Intestinal flora ecology after oral use of antibiotics. Chemother 13:146-160

Drasar BS, Shiner M, McLeod GM (1969) Studies on the intestinal flora. I. The bacterial flora of the gastrointestinal tract in healthy and achlorhydric persons. Gastroenterology 56:71-79

Ellis-Pregler RB, Crabtree C, Lambert HP (1975) The faecal flora of children in the United Kingdom. J Hyg Camb 75:135-142

Evaldson G, Heimdahl A, Kager L, Nord CE (1982) The normal human anaerobic microflora. Scand J Infect Dis, Suppl. 35:9-15

Fernandes CF, Shahani KM, Amer MA (1987) Therapeutic role of dietary lactobacilli and lactobacillic fermented dairy products. FEMS Microbiology Reviews 46:343-356

Finegold SM, Davis A, Miller LG (1967) Comparative effect of broad-spectrum antibiotics on nonspore-forming anaerobes and normal bowel flora. Ann NY Acad Sci 145:268-281

Finegold SM, Sutter VL, Mathisen GE (1983) Normal indigenous intestinal flora. In: Hentges DJ, ed. Human intestinal microflora in health and disease. New York: Academic Press pp 3-31

Fondén R (1989) Lactobacillus acidophilus. In: Les laits fermentés. Actualité de la recherche. London, Paris: John Libbey Eurotext pp 35-40

Gilliland SE (1989) Acidophilus milk products: a review of potential benefits to consumers. J Dairy Sci 72:2483-2494

Goldin BR (1986) In situ bacterial metabolism and colon mutagens. Ann Rev Microbiol 40:367-393

Gorbach S, Barza M, Giuliano M, Jacobus NV (1988) Colonisation resistance of the human intestinal microflora: Testing the hypothesis in normal volunteers. Eur J Clin Microbiol Infect Dis 7:98-102

Gotz V, Romankiewics JA, Moss J, Murray HW (1979) Prophylaxis against ampicillin-associated diarrhea with a lactobacillus preparation. Am J Hosp Pharm 36:754-757

Hall MA, Cole CB, Smith SL, Fuller R, Rolles CJ (1990) Factors influencing the presence of faecal lactobacilli in early infancy. Arch Dis Child 65:185-188

Heimdahl A, Nord CE (1979) Effect of phenoxymethylpenicillin and clindamycin on the oral, throat and faecal microflora of man. Scand J Infect Dis 11:233-242

Heimdahl A, Nord CE (1985) Colonisation of the oropharynx with pathogenic microorganisms - a potential risk factor for infection in compromised patients. Chemioterapia 4:186-191

Hentges DJ (1980) Does diet influence human fecal microflora composition? Nutr Rev 38:329-336

Itoh K, Freter R (1989) Control of *Escherichia coli* populations by a combination of indigenous clostridia and lactobacilli in gnotobiotic mice and continuous-flow cultures. Infect Immun 57:559-565

Kager L, Liljekvist L, Malmborg A-S, Nord CE, Pieper R (1982) Effects of ampicillin plus sulbactam on bowel flora in patients undergoing colorectal surgery. Antimicrob Agents Chemother 22:208-212

Kandler O, Weiss N (1986) Regular, nonsporing Gram-positive rods. In: Sneath PHA, Mair NS, Sharpe ME, Holt JG, eds. Bergey's manual of systematic bacteriology. Vol. 2. Baltimore: Williams & Wilkins pp 1208-1234

Kleeman EG, Klaenhammer TR (1982) Adherence of *Lactobacillus* species to human fetal intestinal cells. J Dairy Sci 65:2063-2069

Knothe H, Dette GA, Shah PM (1985) Impact of injectable cephalosporins on the gastrointestinal microflora: Observations in healthy volunteers and hospitalised patients. Infection 13: S129-133

Lidbeck A, Gustafsson J-Å, Nord CE (1987) Impact of *Lactobacillus acidophilus* supplements on the human oropharyngeal and intestinal microflora. Scand J Infect Dis 19:531-537

Lidbeck A, Edlund C, Gustafsson J-Å, Kager L, Nord CE (1988) Impact of *Lactobacillus acidophilus* on the normal intestinal microflora after administration of two antimicrobial agents. Infection 16:329-336

Macfarlane TW, Samaranayake LP (1989) Clinical oral microbiology. London: Wright

Mackowiak PA (1982) The normal microbial flora. New Engl J Med 307:83-93

Marsh P, Martin M (1984) The normal oral flora. In: Oral microbiology. Berkshire: Van Nostrand Reinhold pp 11-25

Mata L, Urrutia J (1971) Intestinal colonisation of breast-fed children in a rural area of low socioeconomic level. Ann NY Acad Sci pp 93-109

McCarthy C, Snyder ML, Parker RB (1965) The indigenous oral flora of man - I. The newborn to the 1-year-old infant. Arch Oral Biol 10:61-70

Mehta AM, Patel KA, Dave PJ (1983) Isolation and purification of an inhibitory protein from *Lactobacillus acidophilus* AC 1. Microbios 37:37-43

Metchnikoff E (1908) Prolongation of life. New York: GP Putman's Sons

Moore WEC, Cato EP, Holdeman LV (1978) Some current concepts in intestinal bacteriology. Am J Clin Nutr 31: S33-42

Mulligan ME, Citron DM, McNamara BT, Finegold SM (1982) Impact of cefoperazone therapy on fecal flora. Antimicrob Agents Chemother 22:226-230

Nord CE, Kager L (1984) The normal flora of the gastrointestinal tract. Netherlands J Med 27:249-252

Nord CE, Heimdahl A, Kager L (1986) Antimicrobial induced alterations of the human oropharyngeal and intestinal microflora. Scand J Infect Dis Suppl 49:64-72

Nord CE (1990) Studies on the ecological impact of antibiotics. Eur J Clin Microbiol Infect Dis 9:517-518

Oksanen PJ, Salminen S, Saxelin M, Hämäläinen P, Ihantola-Vormisto A, Muurasniemi-Isoviita L, Nikkari S, Oksanen T, Pörsti I, Salminen E, Siitonen S, Stuckey H, Toppila A, Vapaatalo H (1990) Prevention of traveller's diarrhoea by *Lactobacillus* GG. Ann Med 22:53-56

Patte C, Tancrède C, Raibaud P, Ducluzeau R (1979) Premières étapes de la colonisation bacteriénne du tube digestif du noveau-né. Ann Microbiol (Inst Pasteur) 130 A:69-84

Pearce JL, Hamilton JR (1974) Controlled trial of orally administered lactobacilli in acute infantile diarrhea. J Pediatr 84:261-262

Pettersson L, Graf W, Sewelin U (1983) Survival of *L. acidophilus* NCDO 1748 in the human gastrointestinal tract. 2. Ability to pass the stomach and intestine in vivo. In: Hallgren B, ed. Nutrition and the intestinal flora. XV Symp Swed Nutr Found. Uppsala: Almqvist & Wiksell pp 127-130

Rettger LF, Levy MN, Weinstein L, Weiss JE (1935) *Lactobacillus acidophilus* and its therapeutic application. New Haven: Yale University Press

Reuman PD, Duckworth DH, Smith KL, Kagan R, Bucciarelli RL, Ayoub EM (1986) Lack of effect of *Lactobacillus* on gastrointestinal bacterial colonisation in premature infants. Pediatr Infect Dis 5:663-668

Robins-Browne RM, Levine MM (1981) The fate of ingested lactobacilli in the proximal small intestine. Am J Clin Nutr 34:514-519

Roediger WEW (1986) Interrelationship between bacteria and mucosa of the gastrointestinal tract. In: Hill MJ, ed. Microbial metabolism in the digestive tract. Florida: CRC Press pp 201-209

Sandine WE (1979) Roles of lactobacillus in the intestinal tract. J Food Protection 42:259-262

Savage DC (1977) Microbial ecology of the gastrointestinal tract. Ann Rev Microbiol 31:107-133

Savage DC (1979) Introduction to mechanisms of association of indigenous microbes. Am J Clin Nutr 32:113-118

Savage D (1984) Adherence of the normal flora. In: Boedeker EC, ed. Attachment of organisms to the gut mucosa. Florida: CRC Press pp 4-10

Shahani KM, Vakil JR, Kilara A (1977) Natural antibiotic activity of *Lactobacillus acidophilus* and *bulgaricus*. II. Isolation of acidophilin from *L. acidophilus*. Cult Dairy Prod J 12 (2):8-11

Shahani KM (1983) Nutritional impact of lactobacillic fermented foods. In: Hallgren B, ed. Nutrition and the intestinal flora. XV Symp Swed Nutr Found. Uppsala: Almqvist & Wiksell pp 103-111

Silva M, Jacobus NV, Deneke C, Gorbach SL (1987 ) Antimicrobial substance from a human Lactobacillus strain. Antimicrob Agents Chemother 31:1231-1233

Simon GL, Gorbach SL (1981) Intestinal flora in health and disease. In: Johnson LR, ed. Physiology of the gastrointestinal tract. New York: Raven Press pp 1361-1380

Smith RP (1924) The implantation or enrichment of *Bacillus acidophilus* and other organisms in the intestine. Br Med J 11:948-950

Speck ML (1976) Interactions among lactobacilli and man. J Dairy Sci 59:338-343

Speck ML (1983) Lactobacilli as dietary supplements and manifestations of their functions in the intestine. In: Hallgren B, ed. Nutrition and the intestinal flora. XV Symp Swed Nutr Found. Uppsala: Almqvist & Wiksell pp 93-98

Sussman JI, Baron EJ, Goldberg SM, Kaplan MH, Pizzarello RA (1986 ) Clinical manifestations and therapy of *Lactobacillus endocarditis*: Report of a case and review of the literature. Rev Infect Dis 8:771-776

Sutter VL, Finegold SM (1974) The effect of antimicrobial agents on human faecal flora: Studies with cephalexin, cyclacillin and clindamycin. In: Skinner FA, Carr JG, eds. The normal microbial flora of man. Society for Applied Bacteriology, Symposium Series, No 3. New York: Academic Press pp 229-240

Tankanow RM, Ross MB, Ertel IJ, Dickinson DG, McCormick LS, Garfinkel JF (1990) A double-blind, placebo-controlled study of the efficacy of Lactinex in the prophylaxis of amoxicillin-induced diarrhea. DICP Ann Pharmacother 24:382-384

Tannock GW (1983) Effect of dietary and environmental stress on the gastrointestinal microbiota. In: Hentges DJ, ed. Human intestinal microflora in health and disease. New York: Academic Press pp 517-539

Tannock GW, Fuller R, Smith SL, Hall MA (1990) Plasmid profiling of members of the family *Enterobacteriaceae*, lactobacilli, and bifidobacteria to study the transmission of bacteria from mother to infant. J Clin Microbiol 28:1225-1228

van der Waaij D (1983) Colonisation pattern of the digestive tract by potentially pathogenic microorganisms: Colonisation-controlling mechanisms and consequences for antibiotic treatment. Infection 11:90-92

Wells CL, Maddaus MA, Jechorek RP, Simmons RL (1988) Role of intestinal anaerobic bacteria in colonisation resistance. Eur J Clin Microbiol Infect Dis 7:107-113

Wittman DH (1991) Intra-abdominal infections. Pathophysiology and treatment. New York: Marcel Dekker pp 20-29

Yoshioka H, Iseiki K, Fujita K (1983) Development and differences of intestinal flora in the neonatal period in breast-fed and bottle-fed infants. Pediatrics 72:317-321

Yost RL, Gotz VP (1985) Effect of a *Lactobacillus* preparation on the absorption of oral ampicillin. Antimicrob Agents Chemother 28:727-729

Young LS (1988) Antimicrobial prophylaxis in the neutropenic host: lessons of the past and perspectives for the future. Eur J Clin Microbiol Infect Dis 7:93-97

Zoppi G, Deganello A, Benoni G, Saccomani F (1982 ) Oral bacteriotherapy in clinical practice. Eur J Pediatr 139:18-21

Chapter 7

# The Role of Probiotics in the Urogenital Tract

G. Reid, A. W. Bruce and L. Tomeczek

## Introduction

The possibility to treat and prevent infection using non-pathogenic organisms has had appeal for over a century. In particular, the work of Metchnikoff (1894) is recognised for advocating this approach to patient management. In time, his concepts with lactobacilli have led to many investigations into the use of these bacteria for intestinal disorders (Rettger *et al.* 1935). Their potential to prevent infection in the urogenital tract has only been explored in depth in recent years, and this work will form the major focus of this review.

## Concept of Probiotics in Humans

The term probiotics is perhaps more commonly associated with the veterinary field, for example the use of lactobacilli and various microbial cocktails has been described to prevent gastrointestinal infections in chickens, pigs and cows (Snoyenbos *et al.* 1984; Impey *et al.* 1984). In humans, the term bacterial interference has been preferred (Reid *et al.* 1990b), but for the purpose of this article, it will be interchanged with probiosis.

In order to understand the concept of probiotics, it is first necessary to examine the formation and structure of microbial populations in the tissues. Rather than duplicate the content of other contributors to this book, we refer the reader elsewhere for a better understanding of successive acquisition of microflora, colonisation resistance and modes of action.

## Clinical Problem

Two important problems face clinicians in treating a patient with a tissue infection. The first occurs when the primary colonisers have become dominated by the infecting pathogens, either through competition or antibiotic eradication of the indigenous flora. Studies have shown that many pathogens develop dominant adherent microcolonies that become dense biofilms which are able to resist the action of antimicrobial agents (Costerton *et al.* 1987). The option of instilling indigenous flora to challenge these pathogenic biofilms would have a low chance of success due to the succession process which makes it hard for secondary colonisers to outcompete primary ones (Chan *et al.* 1985). The second problem occurs when the pathogens have formed microcolonies within the mucus layers. In an ileal conduit system (urine flowing through intestinal tissue), this type of scenario is very difficult to treat, again due to biofilm resistance to antimicrobial agents and the inability to provide high enough fluid levels to eradicate the organisms (Bruce *et al.* 1984). Other examples could be given, especially where slime producing staphylococci and pseudomonas are involved.

The probiotic technology is not to our knowledge, available to provide an alternative to antimicrobial therapy for these chronic and acute infections. However, as will be discussed later, there may be a place for post-antibiotic replenishment of the flora by instilling indigenous organisms.

# The Genitourinary Tract

The vast array of organisms present within certain tissue sites, especially the genitourinary tract, make it an enormous task to discuss bacterial interference in any great depth. Rather, examples will be cited to illustrate some interesting points, In particular, the discussion will concentrate on the role of lactobacilli.

## Lactobacillus Species

Lactobacilli are rod-shaped Gram positive organisms, which are primarily facultative or strict anaerobes. They are generally non-motile with fastidious growth characteristics that require amino acids, peptides, nucleic acid derivatives, vitamins, salts, fatty acids or fatty esters and fermentable carbohydrates. Thus, their habitat is in many ways determined by access to a complex food supply in an environment with limited or no oxygen. This perhaps explains their presence in the oral, gastrointestinal and genitourinary areas. They are aciduric organisms, prefering a pH of 5.5 - 6.2 and showing poor survival in alkaline conditions. They are speciated according to their fermentation patterns, that is, they are homofermentative (for example *L. bulgaricus*, *L. acidophilus*), facultative heterofermentative (*L. casei*) and obligate heterofermentative (*L. divergens*). Lactobacilli decrease the pH of their environment by forming lactic acid. This acidic environment inhibits the growth of many other species and forms a key part in their competitive activities (Reid *et al.* 1988). These organisms are found in dairy, grain, meat and fish products, beer, wine, fruits, pickled vegetables and other foods. Thus, they are common entrants into the oral and gastrointestinal areas.

For the most part, the mechanisms of action of lactobacillus bacterial interference have not been fully elucidated or confirmed *in vivo*. They would appear to involve a number of factors, as presented in Table 7.1. In short, it seems that the organisms gain an advantage if they colonise as pioneers on the surface, from whence they survive and multiply in a particular niche, competitively exclude pathogens from colonising, inhibit growth of pathogens and form a balanced flora through coaggregation (Reid *et al.* 1987, 1988. 1990b, 1990c; Reid and Bruce 1991).

**Table 7.1.** Characteristics of lactobacilli which are believed to be important in their interference with other organisms in the urogenital tract, for the protection of the host

| |
|---|
| Adhesion - preferably as pioneers or primary successors |
| Growth, multiplication and survival in the mucosal environment |
| Competitive exclusion of pathogenic colonisation |
| Production of inhibitory substances |
| Coaggregation to form a balanced flora |

The activity of inhibitory substances would appear to have at least two mechanisms, namely creation of an acidic environment (Reid et al, 1988) through metabolism of glucose products and production of specific products which inhibit the growth or kill other organisms (McGroarty and Reid 1988a,b). The latter inhibitory mechanism has been exhibited by several lactobacillus isolates, including *Lactobacillus* GG which inhibited a wide range of enteric pathogens (Silva *et al.* 1987), *L. casei* GR-1 which inhibited *E. coli* (McGroarty and Reid 1988a), and *L. fermentum* B-54 which inhibited enterococci (McGroarty and Reid 1988b). The active component of the majority of inhibitory lactobacilli has only been identified to a limited extent, but appears to involve bacteriocins and bacteriocin-like substances.

As mentioned elsewhere in this edition, bacteriocins are proteins or protein complexes with bactericidal activity directed against species that are usually closely related to the producer. They bind to the target cell via relatively specific cell receptors. Often, the substances have a relatively narrow range of action. Bacteriocin-like substances are similar in nature with a narrow spectrum, but they do not inhibit homologous species. It is assumed that these substances are active *in vivo*, but this has not been verified fully. The production of hydrogen peroxide from flavoproteins can also inhibit or kill other organisms, especially if they lack hydrogen scavenging enzymes. The presence of hydrogen peroxide producing lactobacillus has been associated with protection against bacterial vaginosis (Eschenbach *et al.* 1988)

Commercially available *Lactobacillus* products for oral and vaginal insertion, can be obtained from health food stores. They are advertised to coat the gastrointestinal tract, although the scientific basis is weak for these over-the-counter products. The reliability of many of such products has been questioned by the finding that they did not contain the organisms named on the labels (Hughes and Hillier 1990), for poor product viability and for containing strain types foreign to the target site (Hawley *et al.* 1959). There is little evidence to suggest that these preparations have been specifically designed to replenish the mucosal flora and prevent intestinal and/or urogenital infection. Thus, they should be seen for what they are, a food substance of some natural value. Their clinical effectiveness has rarely been subjected to well designed placebo controlled trials, and their use has sometimes been haphazard. However, an understanding of the mechanisms of action of these probiotics will be important to allow their development to full potential.

1a

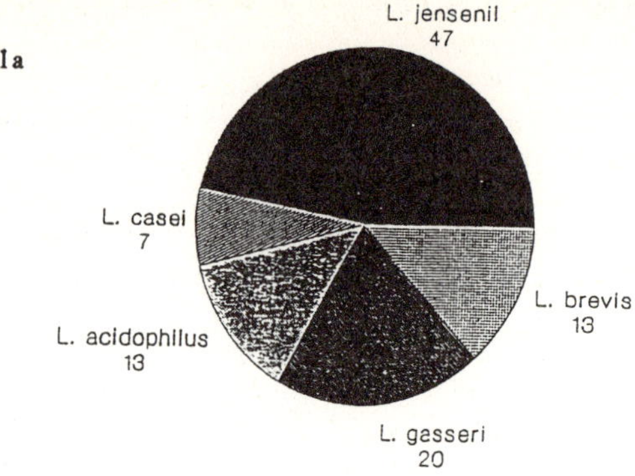


1b

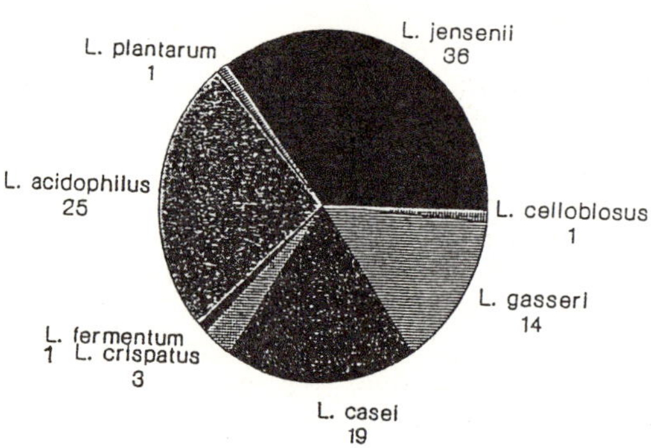


**Fig. 7. 1a,b.** Demonstrates the dominant *Lactobacillus* species isolated from the vagina of 100 healthy adult, premenopausal women (**a**) and from 76 women with a history of recurrent urinary and vaginal infections (**b**). There is no statistical difference in the species distributions.

## Probiotics in the Genitourinary Tract

The urogenital tract is a complex microbiological ecosystem that in the female changes with age, menstrual cycle, pregnancy, time of day and sexual intercourse. One study has been particularly effective at demonstrating hourly changes in flora content and dynamics (Seddon *et al.* 1976). A recent finding has been somewhat surprising, namely that the species of lactobacilli colonising the vagina in healthy patients is similar to those found in patients with a history of infection (Fig. 7.1). This raises several questions. Are lactobacilli not protective against infection? If the answer is that they are protective, the actual numbers of lactobacillus present and their protective properties take on a new significance. It is difficult with current technology and clinical practise to prove that x number of y species of lactobacilli are required to prevent z infection in a patient. However, for the present, we assume that a dominance of lactobacilli is required as these organisms do form the bulk of the normal vaginal flora in healthy women.

Would we have expected a different population in infection prone patients? The answer is no. The lactobacilli found in the urogenital tract have originated from the intestine via intake of food. The patients attending our clinic have access to food sources which meet high western civilisation standards, thus their intestinal flora would not tend to be malnourished or exposed to enteric pathogens from water supplies. While this does not mean that all the intestinal floras will be the same, it is not sufficient to expect vastly different exposures to lactobacilli, In order to study a difference, one could examine patients on dairy and dairy-free diets, then look at lactobacillus presence and types in the genitourinary tract. Whether this causes a difference in protection against urinary tract infection, remains to be seen.

Thus, if we assume that in our patient population, uropathogens encounter the same types of lactobacilli in all the women, is the ability of the pathogens to overcome the flora's competitive activity a virulence factor? Without any concrete experimental evidence, we hypothesise that the pathogens do have a mechanism of overcoming interference factors and establishing themselves in niches, prior to inducing symptoms in the host. The end result in premenopausal women, is that lactobacilli dominate the urethral and vaginal flora of healthy women, whereas during infection and between recurrences, the flora is dominated by pathogens (Bruce *et al.* 1973; Marrie *et al.* 1980). If we were to find out what causes this shift in the ecosystem, we would be better placed to manage the patients.

There is no question that external factors disrupt the flora and alter the equilibrium of the ecosystem. The administration of antibiotics has been found to increase the risk of recurrence of bladder and vaginal infection (Reid *et al.* 1990a; Hertelius *et al.* 1989). Although there is not much data on douches, there are studies to show that nonoxynol-9 in spermicides kills many lactobacilli, while having little or no effect on uropathogens and candida (McGroarty *et al.* 1990a,b). Thus, the use of nonoxynol-9 could increase the risk of acquiring a bladder or vaginal infection.

The ability of lactobacillus to replenish the urogenital flora has been tested *in vivo*. A recent study of ten women showed that weekly insertion of lactobacillus vaginal suppositories led to a significant drop in the bladder infection rate over one year (Bruce *et al.* 1992). This small trial added support to the concept that bacterial interference has applicability in the urogenital tract. But, will this help all patients and is it the optimal approach? It likely will not help every woman with recurrent urinary tract infection as the pathogenesis is a complex process involving many factors. However, for many patients the use of lactobacilli will probably have better compliance and a more natural outcome than antibiotics, presently prescribed on a long term basis. As to the contents of the products, the organisms used in our studies were carefully selected for properties which appeared important to protect the host against infection (Reid *et al.* 1987; McGroarty and Reid 1988b). While additional strains or components could perhaps be added, this product has clear advantages over current commercial preparations.

Another unanswered question is how do we replenish the flora after antimicrobial agents have disrupted it? It has been estimated that one in four women will suffer from a urinary tract infection at some point in their lives, and that most of them will acquire a recurrent infection within one year of their first episode (Kunin 1987). The current therapy for this large patient population comprises three to seven days of an antimicrobial agent. For the most part, this is effective at clearing the bladder. However, seven day therapy can disrupt the urogenital and rectal flora causing drug resistance and uropathogen domination of mucosal tissues (Reid *et al.* 1990a). Even three day therapy with trimethoprim sulfamethoxazole, the most commonly used treatment for bacteriuria, can be followed by a relatively high degree of recurrences (Reid *et al.* 1992).

In Canada, a fluoroquinolone, norfloxacin, is the only drug approved for three day therapy for urinary tract infection. Experience with this drug and other fluoroquinolones suggests that they reach high potency quickly in the bladder, eradicate the pathogens and have minimal effect on the orogenital flora (Reid *et al.* 1990a, 1992). This being the case, one could support the preferred use of three day therapy for simple cystitis as others have recommended (Fair *et al.* 1980). In a study just completed, post-antibiotic administration of lactobacillus vaginal suppositories was found to decrease, to some extent, the recurrence of urinary tract infections (Reid *et al.* 1992). Certainly, the treatment approach had no adverse side effects. A placebo arm using lactobacillus nutrient was not as effective, yet its use raises another conceptual point. While replenishment of lactobacillus flora will often require instillation of viable lactobacillus (many recurrent cystitis patients have fewer if any indigenous lactobacilli), can a second approach involve the stimulation of the growth of the patients own flora? This latter concept could be achieved through the use of special nutritional compounds required by lactobacilli. In a study of 13 healthy women, we found that in all cases, the lactobacillus vaginal flora could be stimulated by insertion of special nutrients (Table 7.2). It is not known if these lactobacilli were potentially protective against uropathogenic challenge. However, there may be some cause to pursue this line of investigation.

**Table 7.2.** The results of lactobacillus vaginal counts after treating 13 women with a single s pecially prepared suppository containing lactobacillus nutrients. In all 13 specimens, the lactobacillus total vaginal count increased by a percentage mean of 81.4±19.7 over one week

| | Lactobacillus viable counts per ml | |
|---|---|---|
| Patient number | Prior to treatment | After treatment |
| 1 | 5 000 | 361 000 |
| 2 | 7 000 | 132 000 |
| 3 | 6 000 | 165 000 |
| 4 | 6 800 | 34 000 |
| 5 | 1 000 | 28 000 |
| 6 | 1800 000 | 2880 000 |
| 7 | 25 300 | 50 000 |
| 8 | 700 000 | 5000 000 |
| 9 | 6 500 | 256 000 |
| 10 | 70 000 | 370 000 |
| 11 | 180 000 | 6200 000 |
| 12 | 4 000 | 9 000 |
| 13 | 2 000 | 17 000 |

## Summary

The difficulty with evaluating probiotics in man, is that a multitude of factors are involved in the selection, use and analysis of outcome for each product used. The science of microbial ecology has still to evolve to the level that demonstrates, with certainty, the mode of action of probiotics. In addition, few groups have been able to develop criteria to select appropriate strains and to employ them in an optimal mode of administration.

At this point in the evolution of the concept, it is apparent that renewed interest in the treatment has come from patients and physicians dissatisfied with current chemical options. There are certainly many examples of human and animal models which support the effective use of bacterial interference, even although unfortunately, few well defined placebo-controlled trials have been carried out. The intestinal and genitourinary tract perhaps provide the best opportunity for further testing of the hypothesis. However, both niches are extremely complex with a wide array of bacteria whose content fluctuates constantly. While the use of genetically manipulated organisms may potentially maximize expression of protective characteristics, the ability of these strains to survive the competition within the host's own flora remains to be tested and proven. Also, the approval of human ethical review boards would need to be obtained for implantation of such mutants. On the other hand nature might prefer to select its own way of replenishing the flora and balancing inserted organisms with indigenous ones.

Clearly the stimulation of non-pathogens to the detriment of actively multiplying virulent pathogens remains a difficult course of action, but one which, without doubt, has important clinical consequences.

## *Acknowledgements*

This work was supported by grants from the Ontario Ministry of Health and the Medical Research Council of Canada. We are grateful to Dr JW Costerton for his input.

# References

Bruce AW, Chadwick P, Hassan A, VanCott GF (1973) Recurrent urethritis in women. Can Med Assoc J 108:973-976

Bruce AW, Reid G, Chan RCY, Costerton JW (1984) Bacterial adherence in the human ileal conduit: a morphological and bacteriological study. J Urol 132:184-188

Bruce AW, Reid G, McGroarty JA, Taylor M, Preston C (1992) Preliminary study on the prevention of recurrent urinary tract infections in ten adult women using intravaginal lactobacilli. Int Urogynecol J 3:22-25

Chan RCY, Reid G, Irvin RT, Bruce AW, Costerton JW (1985) Competitive exclusion of uropathogens from uroepithelial cells by *Lactobacillus* whole cells and cell wall fragments. Infect Immun 47:84-89

Costerton JW, Cheng K-J, Geesey GG, Ladd TI, Nickel JC, Dasgupta M, Marrie TJ (1987) Bacterial biofilms in nature and disease. Ann Rev Microbiol 41:435-464

Eschenbach DA, Davick PR, Williams BL Klebanoff SJ, Young-Smith K, Critchlow CN, Holmes KK (1988) Prevalence of hydrogen peroxide producing *Lactobacillus* species in normal women and women with bacterial vaginosis. J Clin Microbiol 27:251-256

Fair WR, Crane DB, Peterson LJ, Dahmer C, Tague B, Amers W (1980) Three-day treatment of urinary tract infections. J Urol 123:717-721

Hawley HB, Shepherd PA Wheather DM (1959) Factors affecting the implantation of lactobacilli in the intestine. J Appl Bacteriol 22:360-367

Hertelius M, Gorbach SL, Mollby R, Nord CE, Pettersson L, Winberg J (1989) Elimination of vaginal colonization with *Escherichia coli* by administration of indigenous flora. Infect Immun 57:2447-2451

Hughes VL, Hilliar SL (1990) Microbiologic characteristics of *Lactobacillus* products used for colonization of the vagina. Obstet Gynecol 75:244-248

Impey CS, Mead GC, George SM (1984) Evaluation of treatment with defined and undefined mixtures of gut microorganisms for preventing salmonella colonization in chicks and turkey poults. Food Microbiol 1:143-147

Kunin CM (1987) Detection, prevention and management of urinary tract infections. 4th Edition, Lea and Febiger, Philadelphia pp 1-447

Marrie TJ, Swantee CA, Hartlen M (1980) Aerobic and anaerobic urethral flora of healthy females in various physiological age groups and of females with urinary tract infections. J Clin Microbiol 11:654-659

McGroarty JA, Chong S, Reid G, Bruce AW (1990a) Influence of the spermicidal compound nonoxynol-9 on the growth and adhesion of urogenital bacteria *in vitro*. Curr Microbiol 21:219-223

McGroarty JA, Reid G (1988a) Detection of a lactobacillus substance which inhibits *Escherichia coli*. Can J Microbiol 34:974-978

McGroarty JA, Reid G (1988b) Inhibition of enterococci by *Lactobacillus* species *in vitro*. Microb Ecol Health Dis 1:215-219

McGroarty JA, Soboh F, Bruce AW, Reid G (1990b) The spermicidal compound nonoxynol-9 increases adhesion of *Candida* species to human cells *in vitro*. Infect Immun 58:2005-2007

Metchnikoff E (1984) Recherches sur le cholera et les vibrions. IV. Sur l'immunite et la vis-a-vis du cholera intestinal. Ann Inst Pasteur (Paris) 8:529-589

Reid G, Bruce AW (1991) Development of lactobacilli therapy to prevent recurrent urinary tract infections in females. Int Urogynecol 2:40-43

Reid G, Bruce AW, Cook RL, Llano M (1990a) Effect on urogenital flora of antibiotic therapy for urinary tract infection. Scand J Infect Dis 22:43-47

Reid G, Bruce AW, McGroarty JA, Cheng K-J, Costerton JW (1990b) Is there a role for lactobacilli in prevention of urogenital and intestinal infections? Clin Microbiol Rev 3:335-344

Reid G, Bruce AW, Taylor M (1992) Influence of three-day antimicrobial therapy and lactobacillus vaginal suppositories on recurrence of urinary tract infections. Clin Therapeutics 14:11-16

Reid G, Cook RL, Bruce AW (1987) Examination of strains of lactobacilli for properties that may influence bacterial interference in the urinary tract. J Urol 138:330-335

Reid G, McGroarty JA, Angotti R, Cook RL (1988) Lactobacillus inhibitor production against *E. coli* and coaggregation ability with uropathogens. Can J Microbiol 34:344-351

Reid G, McGroarty JA, Dominigue PAG, Chow AW, Bruce AW, Eisen A, Costerton JW (1990c) Coaggregation of urogenital bacteria *in vitro* and *in vivo*. Curr Microbiol 20:47-52

Rettger LF, Levy MN, Weinstein L, Weiss JE (1935) *Lactobacillus acidophilus* and its therapeutic application. Yale University Press, New Haven, Conn

Seddon JM, Bruce AW, Chadwick P, Carter D (1976) Introital bacterial flora - effect of increased frequency of micturition. Br J Urol 48:211-218

Silva M, Jacobus NV, Deneke C, Gorbach SL (1987) Antimicrobial substance from a human *Lactobacillus* strain. Antimicrob Agents Chemother 31:1231-1233

Snoyenbos GH, Weinack OM, Smyser GF (1978) Protecting chicks and poults from salmonella by oral administration of normal gut microflora. Avian Dis 22:273-287

Chapter 8

# Recovery of a Probiotic Organism from Human Faeces after Oral Dosing

S. A. W. Gibson and P. L. Conway

## Introduction

Probiotics have been used therapeutically and prophylactically in bacterial and fungal diseases of humans for a considerable portion of human history. Particularly targeted have been diarrhoea and other manifestations of GI infection (Clements *et al.* 1981; Clements *et al.* 1983; Steffen *et al.* 1986; Salminen *et al.* 1988), oral disease (Perrons and Donoghue 1990) and genitourinary (GU) infection (Will 1979; Sandler 1979; Bruce and Reid 1988; Nagy *et al.* 1991). Probiotics have also been reported to reduce cholesterol levels (Mann and Spoerry 1974; Thompson *et al.* 1982; Jaspers *et al.* 1984), prevent hepatic encephalopathy (Loguercio *et al.* 1987) and degrade carcinogens in the gut (Goldin *et al.* 1980; Cole *et al.* 1989). Results from these studies, however, tend to be contradictory and inconclusive (Pozo-Olano *et al.* 1978; Turnbull *et al.* 1978; Clements *et al.* 1981; Clements *et al.* 1983; Reuman *et al.* 1986). This uncertainty may be due to the use of preparations containing non-viable organisms, the use of microbes from a non-human source, the use of organisms which do not have the properties requisite to function as a probiotic, or to the design of the study. Horikawa (1986) demonstrated in mice that both infected and sterile burn wounds healed significantly faster when an ointment containing a heat killed strain of *Lactobacillus casei* was applied to the affected area and lactobacilli have been used to produce vaccines against unrelated organisms (Gombosova *et al.* 1986).

Anti-viral activity has been attributed to the production by probiotics of hydrogen peroxide (Klebanoff and Belding 1974). Elaboration of antimicrobial substances and lactic acid (Collins and Hardt 1980; Joerger and Klaenhammer 1986; Silva *et al.* 1987; Raccach *et al.* 1989), competition for adhesion sites and substrates (Chan *et al.* 1984; Wilson and Perinin 1988) as well as inactivation of toxins secreted by pathogens (Foster *et al.* 1980) have also been implicated in the mode of action of probiotics. In order for a probiotic to elicit these mechanisms of bacterial interference, the organism must be viable on dosing and survive passage through the gastrointestinal tract.

The GI tract was, however, designed to exclude bacteria from a major opportunity to enter the host system. In the stomach, acid and proteolytic activity destroy bacteria entering via the mouth. In the small intestine microbes are exposed to a cocktail of potentially lethal hydrolytic enzymes and to the powerful antimicrobial effects of bile. On entering the large gut any organism bent on colonisation must come equipped to compete with the established microbial population which occupies every available niche and overcome the considerable colonisation resistance generated by this formidable assembly. Although a number of studies have confirmed the use of preparations containing initially live organisms (Pozo-Olano *et al.* 1978; Gotz *et al.* 1979; Mardh and Soltesz 1983; Grothe *et al.* 1989), too few have attempted to isolate the live probiotic from faeces after oral dosing (Smith 1975; Pattersson *et al.* 1983; Goldin *et al.* 1992).

This chapter describes a logical approach to the isolation and testing of a potential human probiotic. The studies outline steps which take account of:

i the ability of the organism to survive conditions prevalent in the human GI tract;
ii the ability of the organism to produce factors which are antagonistic towards potential pathogens;
iii any unwanted effects produced by the probiotic on the host;
iv the recovery of the organism from faeces during dosing (hence true ability to survive passage through the GI tract);
v the ability of the organism to persist in the gut after dosing has ceased.

# Isolation and Characterisation of the Organism

## Selection

Since lactobacilli are highly host specific (Conway *et al.* 1987; Silva *et al.* 1987), a probiotic for use in humans is more likely to be effective if a human derived *Lactobacillus* strain is used. The bacterium used here was one of several candidate strains isolated from an individual demonstrating life-long resistance to bacterial-type diarrhoeal disease. While this may, at first consideration, appear facile the supporting logic was sound: if the lactobacilli must be selected from a human, one with the requisite properties is more likely to be found in an individual already resistant to GI infection. The bacterium finally selected as the most suitable was identified as *Lactobacillus fermentum* by the API carbohydrate fermentation test and data base with a certainty of 99.9% and designated *L. fermentum* strain KLD.

## Resistance to Acid and Pepsin

Dosing fermented milk-type probiotics to humans is advantageous in that these higher pH, buffered preparations may increase the pH of the stomach to a point where significantly less damage is done to the probiotic organism. The disadvantages (i.e. short shelf life and large volumes required) are, however, sufficiently significant to rule out the use of fermented milk in registered and commercially feasible pharmaceutical products.

In order to avoid expensive and often ineffective acid resistant coatings, the problem of survival in stomach acid must be overcome by selection of an organism which is, *per se*, resistant to pH levels likely to be encountered *in situ*. The organism need not, indeed cannot be expected to, flourish in this environment but must remain viable on exposure to acid for periods of up to 2 h since this is the average time required for stomach emptying.

Bacteria are also sensitive to hydrolytic enzymes, such as proteases, which may cause lysis or damage to proteinaceous surface components (Gibson and Macfarlane 1988; Macfarlane and Cummings 1991). Whilst this latter effect might not result in cell death it may inhibit the growth of the organism sufficiently to compromise persistence in the gut.

In studies conducted with *L. fermentum* KLD no loss in viability was seen during incubation periods of up to 5 h at pH values between 2.5 and 7, however, at pH 1.5 a time dependent decrease in viable numbers was observed (Table 8.1).

**Table 8.1.** Effect of incubation of *L. fermentum* KLD in buffered solutions at a range of pH values

| | $Log_{10}$ viable *L. fermentum* KLD Buffer pH | | | | |
|---|---|---|---|---|---|
| Incubation time | 1.5 | 2.5 | 3.0 | 5.0 | 7.0 |
| 0 | 8.75 | ND | 8.37 | 8.32 | 8.40 |
| 3 | 5.18 | 8.62 | 8.36 | 8.27 | 8.31 |
| 4 | 4.93 | 8.38 | 8.30 | 8.38 | 8.01 |
| 5 | 4.78 | 8.39 | 8.36 | 8.05 | 8.39 |

**Table 8.2.** Effect of incubation of *L. fermentum* KLD in buffered solutions at a range of pH values in the presence of pepsin. The incubation time was 4h

| | $Log_{10}$ viable *L. fermentum* KLD Buffer pH | | |
|---|---|---|---|
| Pepsin Conc (mg/ml) | 1.5 | 2.5 | 4.5 |
| 0 | 4.93 | 8.38 | 8.47 |
| 0.5 | 5.15 | 8.52 | 8.44 |
| 1.0 | 5.39 | 8.45 | 8.42 |
| 2.0 | 6.26 | 8.41 | 8.38 |

Addition of pepsin (0 mg/ml to 2 mg/ml) to the pH 1.5 incubation medium resulted in a concentration dependent increase in survival of the organism (Table 8.2). Reasons for this are unclear but may be due to some protection from low pH afforded by the addition of protein to the medium. Conway *et al.* (1987) demonstrated that bacteria survive better in human gastric juice than in buffer at an equivalent pH indicating that studies using buffers probably underestimate survival potential *in vivo*.

It would appear then that *L. fermentum* KLD was capable of surviving the conditions of low pH and proteolytic activity likely to be encountered in the human stomach.

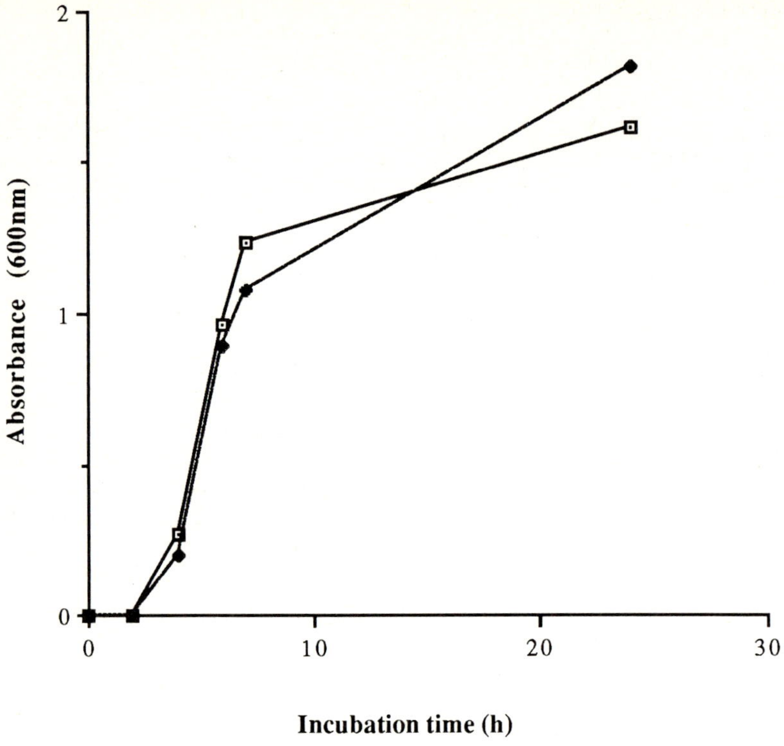


**Fig. 8.1.** Growth of *L. fermentum* KLD in 0% ( □ ) or 0.15% (♦) bile acids.

**Table 8.3.** The growth of *L. fermentum* KLD in MRS broth in the absence or presence of 0.15% oxgall bile. Results are expressed as absorbance (600 nm)

| | Absorbance (600 nm) | |
|---|---|---|
| Incubation time (h) | MRS broth | MRS broth and bile |
| 0 | 0 | 0 |
| 2 | 0 | 0 |
| 4 | 0.27 | 0.20 |
| 6 | 0.96 | 0.89 |
| 7 | 1.23 | 1.07 |
| 24 | 1.62 | 1.82 |

## Survival in Bile

In addition to pH, bile is a powerful antimicrobial defence mechanism in the GI tract.

To determine the susceptibility of *L. fermentum* KLD to bile, actively growing cultures of the organism were added to medium containing 0.15% oxgall bile and incubated at 37°C for 24 h. Samples were removed at intervals for determination of growth using turbidity to estimate cell density (Table 8.3). Although turbidity is not an ideal method for determination of viable numbers the results would appear to demonstrate that *L. fermentum* KLD not only survived but continued to grow actively in the bile medium when compared to control cultures (Fig. 8.1).

## Antagonism to Human Gastrointestinal Pathogens

An important property of any organism intended for use as a probiotic is an ability to be antagonistic towards potential pathogens. This antagonism may consist of production of organic acids, such as lactic acid, or the elaboration of more sophisticated antimicrobials such as bacteriocins. Lactobacilli have been shown to produce both high and low molecular weight antagonistic factors (Tramer 1966; deKlerk and Smit 1967; Upreti and Hindsill 1973; Upreti and Hindsill 1975; McCormick and Savage 1983).

In order to ascertain whether *L. fermentum* KLD secreted any of these antagonistic factors two series of tests were carried out. In the first, two point inoculations (10 ul each) of an actively growing culture of *L. fermentum* KLD were made on BHI agar plates. The plates were incubated for 48 h at 37°C in microaerophilic conditions. Cultures of the human pathogens *E. coli* strains C21, C22, B2C, H10407, 334, R and KS219, *Salmonella sofia* and *Campylobacter jejuni* were cultured for two passages in suitable growth medium and conditions overnight at 37°C. Aliquots of these cultures (10 ul) were mixed with 3 ml of soft BHI agar at 45°C and poured over the *L. fermentum* KLD culture plates. The plates were incubated for a further 24 h. The size of the zone of growth inhibition around the point inoculations of *L. fermentum* KLD were measured and expressed as the distance, in mm, from the edge of the *L. fermentum* KLD colony to the edge of the clear zone.

**Table 8.4.** Antagonistic action of *L. fermentum* KLD against a number of human pathogens is expressed as the size of the clear zone surrounding colonies of *L. fermentum* KLD as measured from the edge of the colony to the edge of the zone, in mm

| Organism | Zone size (mm) |
|---|---|
| *E. coli* C21 | 6.5 |
| *E. coli* C22 | 7.0 |
| *E. coli* B2C | 9.5 |
| *E. coli* H10407 | 7.0 |
| *E. coli* 334 | 8.0 |
| *E. coli* -R | 6.0 |
| *E. coli* KS219 | 5.5 |
| *Salmonella sofia* | 7.0 |
| *Campylobacter jejuni* | 12.0 |

Results (Table 8.4) demonstrated that none of the pathogenic organisms grew over the lactobacilli colonies and that clearing zones of between 5.5 mm and 12 mm were recorded. Organisms most affected were *E. coli* B2C and *C. jejuni*.

These studies do not allow elucidation of the nature of the factor or factors involved in inhibition of growth of the pathogens since lactic acid, present in the agar as a result of fermentation by *L. fermentum* KLD, probably had a significant effect on agar pH and consequently on growth of overlayed cultures. The argument still holds: if these pathogens are inhibited by lactic acid produced *in vitro* may they not similarly be inhibited by lactic acid produced *in vivo*?

The second series of tests were designed to overcome the pH factor. Cultures of *L. fermentum* KLD were grown in BHI broth for 24 h at 37°C under microaerophilic conditions. The bacteria were removed by centrifugation and the spent culture medium was returned to pH 7 using 1M sodium hydroxide. The medium was sterilised by ultrafiltration. Double strength medium suitable for the growth of the pathogenic organisms listed in Table 8.5 was then prepared and sterilised by autoclave. The spent broth was mixed, in equal proportion, with the double strength broth and inoculated with 100 ul of fresh overnight cultures of pathogens. These cultures were incubated in conditions appropriate to the pathogens under test. Growth was assessed by culture turbidity. The effects of factors in the *L. fermentum* KLD spent culture medium were determined by comparison to growth in control cultures. These control cultures were prepared by mixing the appropriate double strength medium with BHI broth which had been incubated, centrifuged and filter sterilised in identical conditions to those of the *L. fermentum* KLD cultures but which had not been inoculated with *L. fermentum* KLD.

**Table 8.5.** Ability of spent culture medium from incubation of *L. fermentum* KLD to inhibit the growth of pathogenic organisms. Results are expressed as per cent reduction in growth compared to control cultures

| Organism | Gram reaction | % reduction |
|---|---|---|
| *Vibreo cholerae* | - | 2 |
| *Shigella* | - | 9 |
| *Pseudomonas aeruginosa* | - | 21 |
| *Escherichia coli* | - | 18 |
| *Enterococcus faecium* | + | 26 |
| *Staphylococcus aureus* | + | 24 |
| *Clostridium perfringens* | + | 27 |
| *Listeria monocytogenes* | + | 36 |
| *Clostridium dificile* | + | 33 |
| *Propionibacterium acnes* | + | 32 |

Results (Table 8.5) showed that most of the organisms were inhibited by the presence of neutralised, spent *L. fermentum* KLD culture medium. The mean decrease in growth rate was 23% (+/-11%) with the maximum inhibition occuring with *Listeria monocytogenese*. Interestingly, but perhaps not surprisingly, there was a statistically significantly greater effect on gram positive organisms than on gram negative organisms ($p<0.05$).

*L. fermentum* KLD, therefore, appeared to have an antagonistic effect on a range of human pathogens *in vitro* through a means not related, at least, to pH. These experiments do not rule out the possibility that the sodium salt of lactic acid had an effect on growth of the pathogens.

## Antibiotic Resistance Profile

One of the more common uses of probiotics in humans may be to overcome the effect on the gastrointestinal microbial ecology had by antibiotic treatment designed to counter a non-related pathogenic state. Studies have shown that treatment with antibiotics can have serious effects on the GI microflora (Heimdahl and Nord 1979; Nord *et al.* 1984) resulting in diarrhoeal disease and, more seriously, pseudomembranous colitis (Aronsson *et al.* 1985; Nord and Edlund 1991). These conditions are, most likely, due to disruption of colonisation resistance. Treatment with probiotics has been shown to off set this loss of protection (Gotz *et al.* 1979; Yost and Gotz 1985) presumably by helping to displace the pathogen responsible and re-establish the normal flora.

For a probiotic to function either as a prophylactic or as a treatment for such conditions, resistance to the antibiotic prescribed would be a distinct advantage. For this reason the resistance of *L. fermentum* KLD to a wide range of antibiotics was tested (Table 8.6).

**Table 8.6.** Antibiotics tested against *L. fermentum* KLD. Results are expressed as manufacturers inhibition values

| Antibiotic | Resistance* |
|---|---|
| Kanamycin | 4 |
| Metranidazole | 4** |
| Novobiosin | 4 |
| Polymixin B | 4 |
| Ceftazidine | 3 |
| Cloxacillin | 3** |
| Flucloxacillin | 3** |
| Gentamycin | 3 |
| Netilmycin | 3 |
| Neomycin | 3 |
| Stroptomycin | 3 |
| Sulfisoxazole | 3** |
| Ampicillin | 2 |
| Cefotaxime | 2 |
| Rifampin | 2 |
| Amoxicillin + Clavulanic acid | 1 |
| Cephaloridine | 1** |
| Clindamycin | 1 |
| Doxycyctin | 1 |
| Erythromycin | 1 |
| Penicillin G | 1 |
| Tetracycline | 1 |
| Chloramphenicol | 0 |

* Resistance: 4-highly resistant, 3-very resistant, 2-resistant, 1-minimal resistance, 0-sensitive

** Estimated from inhibition value and concentration

*L fermentum* KLD was grown overnight in MRS broth at 37°C in microaerophilic conditions and aliquots used to evenly inoculate MRS agar plates. Commercially available antibiotic impregnated discs were placed on the agar surface and the plates incubated as described for 48 h. Plates were uniformly prepared to enable direct comparison of results between samples.

**Table 8.7.** Antibiotic resistance profile

| Antibiotic group | Resistance |
|---|---|
| Penicillins | ++ |
| Cephalosporins | +++ |
| Tetracyclines | + |
| Aminoglycosides | ++++ |
| Macrolides | + |
| | |
| Others | |
| Chloramphenicol | - |
| Polymixin B | ++++ |
| Rifampin | ++ |
| Novobiosin | ++++ |
| Metronidazole | ++++ |
| Sulphisoxazole | +++ |

Antibiotic resistance was quantitated by measuring the size of the zone of non-growth around each disc and relating this to resistance using the manufacturers guidelines. The results are summarised in Table 8.7.

As can be seen the *L. fermentum* KLD was resistant to a number of major antibiotic classes, such as aminoglycosides and cephalosporins but sufficiently susceptible to, for example, certain of the tetracyclines and macrolides, to ensure that the organism could be eradicated if necessary.

## Survival in Competition with Human Faecal Bacteria

The ability of *L. fermentum* KLD to compete with other organisms present in human faeces was tested *in vitro* using a single chamber chemostat after the method of Allison *et al.* (1989). The chemostat was inoculated with a 20% slurry of human faeces prepared in sterile anaerobic buffer. The dilution rate was 0.085/h and the culture was allowed to equilibrate for four days before inoculation with *L. fermentum* KLD. After inoculation the dilution rates and chemostat sampling was as shown in Table 8.8. The identification of *L. fermentum* KLD was carried out using selective growth medium, growth conditions and plasmid profiling as described.

Results demonstrated that *L. fermentum* KLD maintained a population of between $10^{10}$ and $10^{11}$ cfu per ml at dilution rates of 0.042/h and 0.15/h. This faster rate was selected to imitate conditions in the GI tract during the fast and volumous transit seen in diarrhoea. It would appear from these studies that, at least *in vitro*, *L. fermentum* KLD was able to persist in the presence of other human faecal bacteria even at high dilution rates.

**Table 8.8.** Growth of *L. fermentum* KLD in chemostat culture with a mixed pupulation of human faecal bacteria

| Days after inoculation | Dilution rate* | Viable count** |
|---|---|---|
| 0 | 0.085 | $6 \times 10^8$ |
| 1 | 0.085 | $3 \times 10^8$ |
| 7 | 0.042 | $7 \times 10^{10}$ |
| 8 | 0.042 | $2 \times 10^{10}$ |
| 9 | 0.042 | $1 \times 10^{10}$ |
| 10 | 0.150 | $1 \times 10^{10}$ |
| 13 | 0.150 | $5 \times 10^{10}$ |
| 14 | 0.150 | $7 \times 10^{10}$ |

* Dilution rate in $L^{-1}$ (D=0.15 is equivalent to 1.8 $ld^{-1}$)
** Colony forming units $ml^{-1}$

# Phase 1 Studies

After completion of the pre-clinical studies described above three main studies were left to be done:

i production of *L. fermentum* KLD in commercial quantities and suitable formulation;
ii stability studies on the selected formulation; and
iii Phase 1 studies in non-patient human volunteers.

It is beyond the scope of this chapter to describe production of bacteria on a commercial scale or to detail stability studies required subsequently. These topics are covered in Chapter 10 this volume.

Large scale fermentation was carried out and the batches freeze-dried resulting in a powder containing around $10^{12}$ cfu per gram. The powder was encapsulated in size 00 clear, hard gelatin capsules containing between $1 \times 10^{11}$ and $3 \times 10^{11}$ cfu.

Stability studies showed that at temperatures up to 20°C no loss of viability occurred over the period tested. At 28°C some loss occurred and at temperatures in excess of this rapid loss in viability occurred. As well as viability, tests were carried out to ensure that no loss in resistance to bile acid, stomach acid, proteolytic activity and a selection of representative antibiotics had occurred. Plasmid profiles were also used to check strain purity. These studies continued for a period of 18 months.

## Study 1

A pilot study to determine the persistence of *L. fermentum* KLD in the gastrointestinal tract of healthy non-human volunteers following a single dose of $10^{11}$ viable organisms.

## Summary of Study Design

This was a single dose pilot study to investigate the persistence of *L. fermentum* KLD in the GI tract of health, non-patient volunteers after a single dose of around $1 \times 10^{11}$ cfu.

The protocol was approved by a local, independent Ethics Committee and the study was conducted in accordance with the 'Declaration of Helsinki'. Each volunteer gave written informed consent.

Six male volunteers between the ages of 24 and 44 years, who were within +/-15% of the weight limits defined by the Metropolitan height/mass tables were selected. Each volunteer underwent a full medical, physical and biochemical assessment prior to the study. All were considered normal by the study physician. Exclusion criteria included presence of GI, cardiovascular, hepatic or renal disease, treatment with antibiotics within three months of the start of the study and major surgery or medical illness in the three months before the study. Each volunteer provided a faecal sample within two weeks of the start of the study to confirm the absence of lactobacilli with a plasmid profile identical to that of *L. fermentum* KLD.

Within the week before the start of the study each volunteer underwent a medical examination which included heart rate, height, weight, blood pressure, respiratory rate, temperature and a general examination. An ECG recording was also made and a full medical history taken.

A 14 ml sample of venous blood was taken and assayed for parameters of haematology and blood biochemistry. Urinalysis was performed at the medical examination.

The *L. fermentum* KLD was administered on the morning of the first day of the study after a fasting period lasting from 2200 h the previous night and a standard breakfast. The volunteers were observed for 30 min after the dose then allowed to continue their normal routine. All volunteers were required to return to the study base on days 2, 3, 4, 5, 8, 15, 22 and 29, when they were questioned on any adverse events and about their health in general. They completed a short questionnaire on gastrointestinal symptoms experienced using a visual analog scale (VAS) to quantify their responses.

Volunteers collected all faeces voided on days 1 to 7 and on days 11, 15, 22 and 29. Further samples were collected if required and three consecutive samples completely free of *L. fermentum* KLD were required to prove no further persistence of the organism in the volunteers. The faecal samples were stored in an incubator at 37°C overnight and assayed for the presence of *L. fermentum* KLD the following day.

## Isolation of Lactobacilli

The 24 h faecal output of each volunteer was pooled and mixed to homogeneity. An aliquot (1 g) was dispersed in anaerobic half-strength de Mann Sharp Rogosa (MRS) broth and decimally diluted (in duplicate) in the same medium. Relevant dilutions were inoculated on to Rogosa agar and the plates incubated for 48 h at 37°C in an atmosphere of 10% $CO_2$ in air microaerophilic (conditions). Colonies were selected for further analysis on the basis of colony morphology and microscopic appearance. These colonies were inoculated into MRS broth and incubated overnight as described. Gas production was noted as characteristic of *L. fermentum* KLD but was not used to disqualify non-producers at this stage.

The overnight cultures were checked for purity microscopically and either used immediately for plasmid analysis or stored frozen at -70C in sterilised milk until required.

## Plasmid Analysis

The fresh overnight cultures or freshly thawed frozen samples were inoculated into 30 ml of MRS broth and incubated at 37°C for 6 h. The exponentially growing cells were subject to plasmid extraction and analysis by a modification of the method of Tannock *et al.* (1990). The bacteria were washed and resuspended in tris buffer (pH 8.2). Lysozyme (6 mg/ml) was added and the mixture incubated for 25 min at 37°C. At this time sodium dodecal sulphate (SDS) and sodium hydroxide were added. Lysed cell suspensions were neutralised by the addition of concentrated tris buffer (pH 7.2) and the cell debris precipitated with 5M sodium chloride. After RNA digestion, the remains were extracted with a phenol and chloroform mixture and the aqueous phase cold precipitated with sodium acetate. The precipitate was redisolved in tris buffer (pH 8.5) and nucleic acids further purified and concentrated using ion exchange resin. The resultant sample was subjected to electrophoresis in 0.7% agarose at 30 mA for approximately 2 h. Plasmid bands were visualised through UV transillumination after staining with ethidium bromide. A permanent record was made using a Polaroid DS-34 direct screen camera.

## Calculation of the Number of *L. fermentum* KLD in Faecal Samples

After selection of colonies from faecal samples as described above definitive identification was obtained by comparison of the plasmid profiles of the selected colonies with that of authentic *L. fermentum* KLD. The actual number present per gram of faeces was calculated in the standard way from the number of colonies identified as *L. fermentum* KLD by the criteria described.

## Effect of Overnight Incubation at 37°C on *L. fermentum* KLD Populations

To test the effect of the storage of faecal samples at 37°C overnight human faeces were collected from volunteers who had taken part in the study. These samples were diluted 10- or 100- fold in half strength, anaerobic saline to facilitate even dispersion of added bacteria. Half strength saline was used since it would not stimulate bacterial growth. The faecal suspensions were inoculated with *L. fermentum* KLD to give final counts of approximately $10^3$, $10^4$, $10^5$ and $10^6$ cfu per 10 ml of suspension. Actual numbers were determined by immediately plating diluted aliquots of these suspensions, in duplicate, on Rogosa agar. The plates were incubated in microaerophilic conditions for 48 h before counting. The remainders of the faecal suspensions were incubated in air-tight containers overnight at 37°C. After this time the number of lactobacilli were determined as described. Volunteers with low levels of commensal microaerophilic organisms were utilised to allow determination of *L. fermentum* KLD by morphology and microscopic appearance only.

Results showed that there was a small increase in the number of *L. fermentum* KLD present in the samples after overnight incubation (Table 8.9). Linear regression analysis of the number of organisms recovered against the number added gave a correlation coefficient of 0.959. The difference between counts added and counts recovered was not statistically significant at the 5% level (n=0.09). On average the number of *L. fermentum* KLD recovered after the incubation was 1.46-fold higher than control counts. It is interesting and perhaps relevant here to consider that if, in general terms, this were a real increase in numbers, only 46% of the *L. fermentum* KLD present divided to produce daughter cells. In contrast, *L. fermentum* KLD under ideal conditions would divide every 30 min- 40 min, producing and increase in numbers of greater than $10^{10}$ -fold in 24 h! This small study showed that under the storage conditions defined here, and used in the study, *L. fermentum* KLD survived and gave good recoveries. It is highly likely, therefore, that estimates of numbers are reliable and accurate.

**Table 8.9.** Number of *L. Fermentum* KLD before and after overnight incubation at 37°C

| Sample | No organisms added* | No organisms recovered |
|---|---|---|
| 1 | 2000 | 3100 |
| 2 | 6200 | 8350 |
| 3 | 560 | 850 |
| 4 | 560 | 800 |
| 5 | 4250 | 6700 |
| 6 | 4250 | 4600 |
| 7 | 4250 | 7400 |

* cfu ml-1

# Results of the Single Dose Study

## Clinical Findings

Analysis of clinical findings before, during and after the study indicated that none of the volunteers showed any adverse event or reaction associated with the medication.

The results of the VAS assessment demonstrated small but statistically significant improvement in upper abdominal discomfort after the medication. Small, statistically non-significant improvements in lower abdominal discomfort and flatulence were noted. Average bowel function and general well-being were unaffected. Any trends in these parameters were treated with caution due to the open design of the study and the small number of volunteers participating. They do confirm, however, that no serious adverse events occured after ingestion of the medication.

## Faecal Output

Faecal samples collected from each volunteer were weighed to allow excretion of *L. fermentum* KLD to be calculated. These data (Table 8.10) showed that faecal output varied from 23.6 g/day to 368.9 g/day. The mean faecal output was 134.3 +/- 79.6 g/day (Mean +/- SD). This variation occurred without undue concern from the volunteers and was, presumably considered normal by them. Interestingly, one volunteer reported reduced faecal output (constipation) on day two, however examination of faecal mass voided on days two and three for this individual (247.4 g and 203.3 g respectively) did not support this claim since the mean faecal output during the entire study was 215.6 +/- 83.9 g for this volunteer. The mean faecal output for all volunteers was higher than that previously reported (Cummings *et al.* 1990) although comparisons were made difficult by composition of the groups. A total of 66 samples were available for consideration in this study.

**Table 8.10.** Faecal output (g) for volunteers 1-6 during the Single Dose study

| | Volunteer | | | | | |
|---|---|---|---|---|---|---|
| Day | 1 | 2 | 3 | 4 | 5 | 6 |
| 1 | 72.7 | 63.0 | MS | 240.6 | 96.9 | 368.9 |
| 2 | 121.5 | 204.6 | 151.1 | 91.3 | 89.8 | 247.7 |
| 3 | 63.3 | 79.3 | 131.7 | 88.3 | 59.9 | 203.3 |
| 4 | 121.0 | 0.0 | 132.5 | 140.9 | 59.3 | 286.6 |
| 5 | 97.3 | 67.4 | 158.7 | 59.0 | 48.9 | 321.0 |
| 6 | 53.2 | 207.1 | 114.6 | 48.7 | 95.8 | 133.3 |
| 7 | 162.5 | 144.1 | 48.9 | 89.0 | 198.0 | 146.7 |
| 11 | 72.7 | 215.3 | 105.9 | 28.1 | 34.9 | 220.5 |
| 15 | 75.8 | 23.6 | 85.7 | 320.3 | 273.1 | 91.3 |
| 22 | 0.0 | 170.7 | 101.9 | 141.6 | 0.0 | 233.7 |
| 29 | 87.6 | 268.3 | 93.9 | 148.8 | 105.8 | 119.0 |
| 36 | - | - | 95.9 | - | 175.7 | - |
| Mean | 84.3 | 131.2 | 110.9 | 127.0 | 103.2 | 215.6 |
| ±SD | 40.3 | 84.7 | 31.5 | 83.1 | 73.9 | 83.9 |

- No sample collected
MS Missed sample

## Excretion of *L. fermentum* KLD

All volunteers in the study excreted *L. fermentum* KLD at some time during the collection period (Table 8.11) indicating that pre-clinical assessment of the ability of the organism to survive passage through the human GI tract was valid here.

**Table 8.11.** Total faecal excretion (log 10 cfu) of *L. fermentum* KLD by volunteers 1-6 during the Single Dose Study

| | Volunteer | | | | | |
|---|---|---|---|---|---|---|
| Day | 1 | 2 | 3 | 4 | 5 | 6 |
| 1 | 11.0 | 0 | 0 | 8.8 | 0 | 0 |
| 2 | 11.5 | 7.6 | 11.3 | 8.6 | 7.4 | 5.1 |
| 3 | 13.0 | 0 | 8.7 | 8.3 | 0 | 0 |
| 4 | 8.8 | 0 | 10.3 | 0 | 0 | 0 |
| 5 | 8.6 | 0 | 10.4 | 6.5 | 6.9 | 0 |
| 6 | 8.2 | 0 | 8.1 | 0 | 6.6 | 0 |
| 7 | 0 | 0 | 0 | 0 | 9.5 | 0 |
| 11 | 0 | 0 | 0 | 8.6 | 8.7 | 0 |
| 15 | 0 | 0 | 0 | 0 | 9.9 | 0 |
| 22 | 0 | 0 | 0 | 0 | 0 | 0 |
| 29 | 0 | 0 | 0 | 0 | 10.2 | 0 |
| 36 | - | - | 0 | - | 11.2 | - |

It is possible that some organisms may survive the *in vitro* tests but fail to persist in the GI tract or, conversely, that some may fail to survive *in vitro* but thrive in the gut. It would appear, however, that the *in vitro* tests used here may present a screen for the selection of potential probiotic organisms.

Volunteers excreted *L. fermentum* KLD in three distinct patterns. Two volunteers showed a pattern characterised by the appearance of the bacterium in faeces at 48 h, then absence from subsequent samples. In three individuals *L. fermentum* KLD was present in faeces from 24 h to between six and 11 days after dosing. The remaining volunteer continued to excrete the organism for up to 36 days.

With such small numbers of volunteers it is dangerous to speculate on the meaning or relevance of these patters. Larger studies may indicate the presence of other excretion patterns or discredit the validity of those found here.

It is possible, however, that these patterns represent three distinct types of reaction to ingestion of *L. fermentum* KLD. Individuals who excrete the organism for only one day may have physiological, immunological or microbiological barriers to long-term persistence of orally dosed, non-commensal organisms. Those who excreted for between six and 11 days may have more tolerance, with rejection factors taking longer to act on the extraneous bacterium. The individual who excreted for 36 days may be particularly tolerant to non-commensals or harbour native organisms which closely resembled *L. fermentum* KLD in some factor which permitted such survival.

Total excretion of *L. fermentum* KLD varied from $10^5$ to $10^{13}$ organisms. Volunteers did not excrete significantly more of the organism than was given in the dose. It is possible that the bacterium merely passed through the gut and was recovered from faeces. Given the conditions in the gut, however, and especially those in the large gut, it is likely that growth and cell division are necessary for detection of even small numbers in faeces.

The overall purpose of this study was to examine the safety of *L. fermentum* KLD in human volunteers and to ascertain whether the organism could be recovered with certainty of identification in faeces. Results confirm that no unwanted effects were observed and that the organism could be recovered in faeces after oral dose.

## Study 2

A study to determine the growth and persistence of *L. fermentum* KLD in the gastrointestinal tract of health, non-patient volunteers following administration of doses of either $10^8$ or $10^{11}$ organisms twice daily for fourteen days.

### Summary of Study Design

This was randomised, parallel, multiple dose study in 12 healthy, non-patient volunteers. Faeces were collected from the volunteers on days 1-14 to establish background levels of commensal microaerophilic lactobacilli and to ensure the absence of organisms with a similar plasmid profile to *L. fermentum* KLD. After this pre-screening period, capsules containing either $10^8$ or $10^{11}$ organisms were self-administered for a further 14 days (days 15-28).

The persistence and population levels of *L. fermentum* KLD in faeces from the volunteers were assessed before administration (days 1-14), during administration (days 15-28) and after dosing ceased (days 29-38, 44-46 and 57).

A range of relevant statistical methods were used to compare results between groups and between study periods.

The 12 volunteers were selected as described in Study 1 and Study 2 was conducted under the same conditions of written, informed consent and the "Declaration of Helsinki": medical monitoring was extensive. VAS scoring systems were again used.

## Study Medication

Volunteers received one of the two treatments listed below, according to a randomisation list (Table 8.12).

**Table 8.12.** Randomisation list used to assign volunteers in the Multiple Dose Study to high or low doses

| Volunteer | 1 | 2 | 3 | 4 | 5 | 6 | 7 | 8 | 9 | 10 | 11 | 12 |
|---|---|---|---|---|---|---|---|---|---|---|---|---|
| Treatment | B | B | A | A | A | B | B | A | A | B | A | B |

A Low dose ($10^8$ L. fermentum/day)
B High dose ($10^{11}$ L. fermentum/day)

*Treatment A* Volunteers assigned to this treatment received 28 gelatin capsules containing between $1 \times 10^8$ and $5 \times 10^8$ viable *L. fermentum* KLD. The capsules were swallowed whole with water twice daily, morning and evening, after food for a period of 14 days.

*Treatment B* The six volunteers in this group received 28 gelatin capsules containing between $1 \times 10^{11}$ and $5 \times 10^{11}$ viable organisms. Capsules were administered as described for Treatment A.

These levels were used in an attempt to establish a 'non-effective' dose of *L. fermentum* KLD. The lower dose was selected due to ease of production of material containing the correct level of the organism and based on published work indicating that, for a probiotic similar to *L. fermentum* KLD, detection in faeces did not occur unless the dose was greater that $10^9$ viable organisms (Saxelin *et al.* 1991).

# Results from the Multiple Dose Study

## Clinical Findings

Clinical assessments made throughout the study showed a number of differences between the two treatment groups although at no time did any of these reach statistical significance. During the dosing period an increase in lower abdominal discomfort and flatulence was observed in the higher dose group and an increase in upper abdominal discomfort was indicated for both groups; none of these was statistically significant.

**Table 8.13.** Mean faecal output (g) for all volunteers in the Multiple Dose Study

| | Volunteer | | | | | | | | | | | |
|---|---|---|---|---|---|---|---|---|---|---|---|---|
| Trial period | 1<br>H* | 2<br>H | 3<br>L | 4<br>L | 5<br>L | 6<br>H | 7<br>H | 8<br>L | 9<br>L | 10<br>H | 11<br>L | 12<br>H |
| Pre-dosing | 168.7<br>(46.4) | 223.7<br>(95.2) | 181.5<br>(98.6) | 189.7<br>(85.2) | 134.7<br>(63.2) | 123.1<br>(42.9) | 217.6<br>(63.7) | 166.2<br>(97.8) | 78.6<br>(44.1 | 141.5<br>(72.8) | 74.7<br>(53.9) | 109.2<br>(77.6) |
| Dosing | 189.1<br>(49.6) | 239.6<br>(96.4) | 186.2<br>(98.9) | 192.4<br>(124.2) | 119.2<br>(62.7) | 134.7<br>(39.4) | 161.3<br>(68.4) | 124.9<br>(68.4) | 72.7<br>(64.3) | 133.5<br>(60.8) | 109.0<br>(73.3) | 110.0<br>(83.9) |
| Post-dosing | 153.1<br>(54.7) | 225.7<br>(102.3) | 158.1<br>(121.4) | 129.5<br>(90.1) | 108.4<br>(76.0) | 130.6<br>(46.0) | 217.5<br>(95.5) | 128.3<br>(54.7) | 115.3<br>(58.2) | 138.0<br>(58.2) | 112.3<br>(80.2) | 121.9<br>(84.0) |

*H - high dose, L - low dose
Numbers in parenthesis are standard deviations

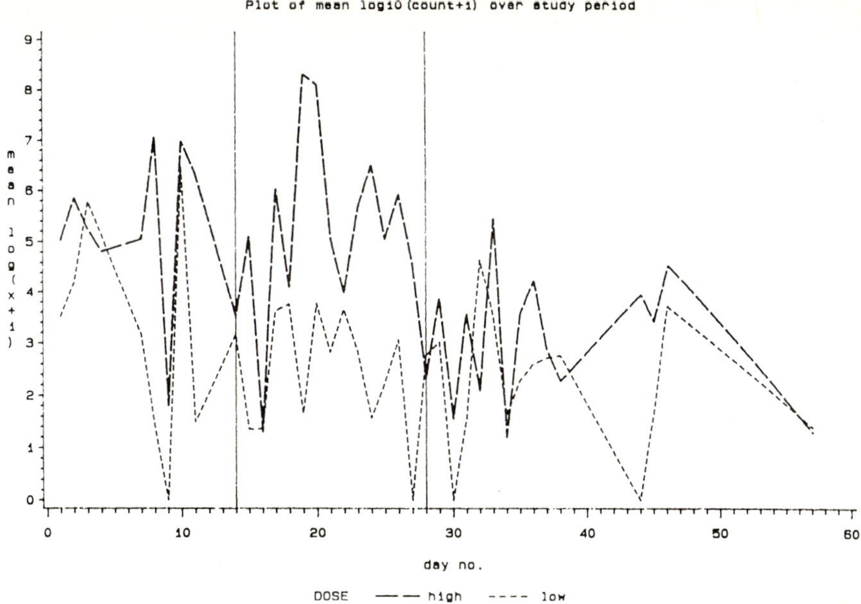


**Figure 8.2.** Total microaerophilic lactobacilli excreted per day.

Bowel function changed during the dosing period for the low dose group from normal to looser faeces. None of the effects was severe and were not statistically or clinically significant nor were they related to the treatment material.

There was, as with the single dose study, a slight but non-significant improvement in general well-being.

There were 40 unwanted effects reported over the treatment period, however none of these was considered attributable to the probiotic. None of the volunteers was withdrawn from the study. Adherence to the protocol was good and the medication was well tolerated.

## Faecal Output

Mean faecal output for each volunteer during each period of the study is shown in Table 8.13. Analysis of results showed some indication of an increase in faecal output for both low and high dose groups during the early part of the dosing period when compared to the pre-dosing period. The effect was on the verge of statistical significance at the 10% level with an increase of 27.5 g for the high dose group and 20.3 g for the low dose group. Output had returned to pre-dosing levels by day 23 and no other significant changes were observed. There was no statistically significant difference in faecal output between the two groups.

The overall mean faecal output for all volunteers was 148 g/day and was not different from that recorded in the single dose study. It might be interesting here to consider that 12 volunteers each produced on average 148 g of faeces per day for a total of 42 days resulting in a problem massing some 75 k for the investigators to solve!

## Excretion of Microaerophilic Lactobacilli

Total excretion of microaerophilic organisms able to grow on Rogosa agar (hereafter referred to as microaerophilic lactobacilli (ML) was determined for each volunteer on each day of the study. Individual data for the 14 day period before the start of the study showed considerable intra- and inter-individual variation in the numbers of ML excreted. Mean data for the Low and High dose groups are shown in Figure 8.2. No statistically significant differences in excretion were detected between the groups during this pre-dosing period.

During the dosing period volunteers receiving the high dose of *L. fermentum* KLD excreted significantly more ML than those on the low dose indicating that dosing with $10^{11}$ *L. fermentum* KLD produced an increase in the excretion of this type of organism.

In the period after dosing ceased both groups again excreted similar numbers of ML. Return to the pre-dosing pattern of excretion occurred within 48 h of cessation of ingestion of *L. fermentum* KLD.

No significant difference in the numbers of ML excreted for either volunteer group was observed when the pre- and post-dosing periods were compared.

From these data it would appear that dosing $10^{11}$ *L. fermentum* KLD significantly increased the numbers of ML recovered in faeces but, however tempting it is to conclude that these organisms are *L. fermentum* KLD, such a conclusion is unsafe. It is possible that dosing *L. fermentum* KLD in some way stimulated an increase in numbers of a commensal lactobacilli or lactobacilli generally.

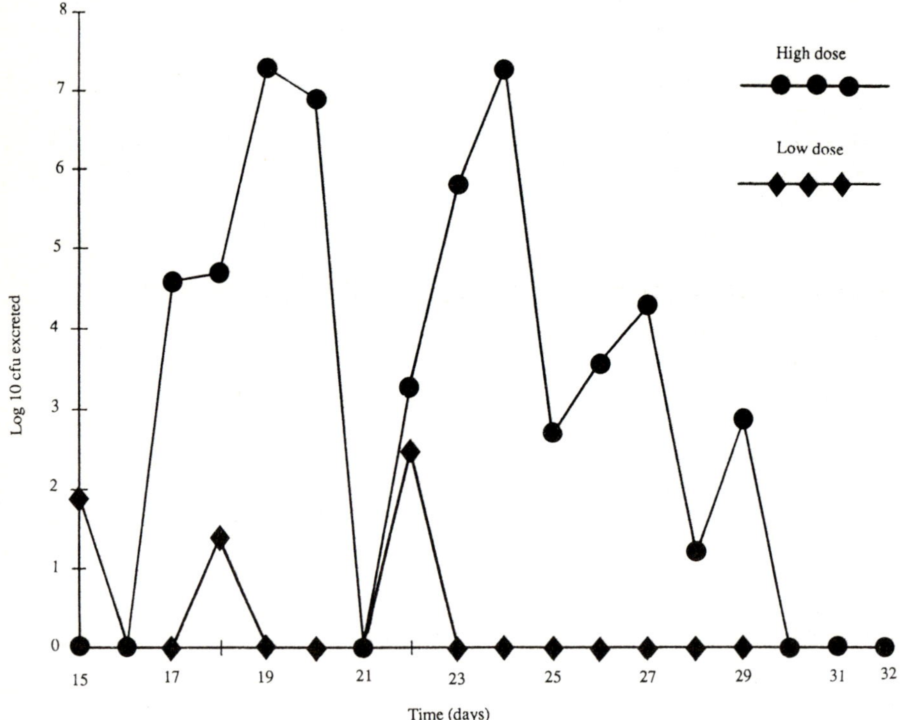


**Figure 8.3.** Excretion of *L. fermentum* KLD by volunteers on high or low dose.

To rule out such possibilities, plasmid profiling was used to confirm the presence of *L. fermentum* KLD, specifically. Results confirmed that the increase in numbers of ML during dosing of $10^{11}$ cfu was indeed due to the presence of *L. fermentum* KLD (Fig. 8.3)

## Observations from the Multiple Dose Study

Excretion of ML varied tremendously both between individuals and from day to day in the same individual. All feasible precautions were taken to minimise variation in growth conditions and handling procedures. Bacterial media were bought in sufficient quantity for the entire study from the same commercial batch and prepared fresh every day. Despite this, excreted numbers of ML could vary from between 0 cfu and $10^{10}$ cfu in one volunteer from one day to the next.

Reasons for this variation may be guessed at. Conditions in the gut may vary widely. Dietary components may affect growth of these organisms or dietary intake of microaerophiles may be reflected in faecal output. It is possible that these bacteria colonise the surface of the gut forming colonies, perhaps as biofilms, which are periodically sloughed from the mucosa causing the wide variation observed here. Unfortunately, it was beyond the scope of this study to investigate these possibilities.

In the single dose study, three patterns of excretion of *L. fermentum* KLD were observed in volunteers receiving a single dose of the organism. In this study all volunteers on the high dose of *L. fermentum* KLD had ceased excreting the organism by 48 h after the last dose. Here again further work is required to determine the cause of this abrupt end to excretion.

Since there were only two dosage levels given in this study a minimum dose producing significant levels of faecal excretion of *L. fermentum* KLD was not established. The results did establish that a dose of $10^8$ or less did not lead to detection of the organism in faeces. This is in broad agreement with previously published results (Saxlin *et al.* 1991).

## Safety Issues Related to the Use of Bacteria in Humans

The administration of live bacteria to humans raises certain safety issues in the eyes of the Medicines Control Agency (MCA), despite the fact that fermented foods, yoghurrts and unlicenced probiotics have been consumed by man for generations.

In order to help assuage these concerns, the role of lactobacilli in pathogenicity in man and the potential of transfer of plasmids containing pathogenicity factors to lactobacilli from other microorganisms in the gut were assessed.

## Pathogenicity of Lactobacilli in Man

Review of the literature revealed that pathological conditions associated with lactobacilli in man are rare. Where cases have occurred pre-existing disease states were present.

## Endocarditis

Up to the middle of 1988 only 24 cases of endocarditis in which lactobacilli were present had been reported (Naudé *et al.* 1988)). In 83% of these cases pre-existing structural heart disease was present, and an identifiable pre-disposing event was present in 79%. These included poor dentition (eight cases), tooth extraction (five cases), dental abscess (two cases), dental scraping (one case) maxillary ostetitis (one case) and parturition (one case). In patients where no existing heart disease was noted, recent dental work had been conducted.

## Lactobacillaemia

By describing a further nine cases, Bayer *et al.* (1978) increased the number of world-wide reported accounts of bacteraemia caused by lactobacilli to 24. In six of the nine individuals, endocarditis was not present. Three of these had localised abscesses, two in the peritomeum following enteric perforation and one in the oral cavity. The remaining cases were two individuals with postpartum endometritis and a premature neonate with peri-omphalitis. Other predisposing factors were intravenous drug abuse, previous streptococcal endocariditis and carious teeth. Identification to species level was possible in six cases, the organisms being *L. leichmanii*, *L. acidophilus* (two cases), *L. plantarum* (two cases) and *L. casei*.

## Oral Pathology

Lactobacilli comprise a small but characteristic group of oral bacteria (Naudé *et al.* 1988). Early findings implicated lactobacilli in the formation of dental caries since they were present (albeit in low numbers) on the tooth surface before and after lesion formation, and were both aciduric and acidogenic (Howe and Hatch 1917). More recently, however, Streptococcus mutans has been shown to be the cariogenic organism in both humans (Loesche and Straffon 1979) and experimental animals (Michalek *et al.* 1981). Studies in germ-free rats (Michalek *et al.* 1981) demonstrated that *L. casei* associated mainly with the soft oral tissues and mucous membranes. Compared to *S. mutans*, *L. casei* only poorly colonised tooth surfaces and, instead of contributing to pathogenesis appeared to protect the tooth from *S. mutans* induced lesion formation.

## Other Infections

One case of urinary tract infection caused by *L. gasseri* was reported in a 66 year old man with diabetes and generalised atherosclerosis (Dickgieser *et al.* 1984). The organism was probably present as an opportunist.

In another case *L. acidophilus* was isolated from the oesophagus of a 67 year old woman. The patient had a mitral valvulotomy and replacement of the aortic valve. A long course of tetracycline had been given and the patient suffered from retrosternal burning and dysphagia. *L. acidophilus* was isolated from an osophogeal biopsy but further antibiotic treatment had no effect and the patient died of cardiac failure, probably due to the earlier surgery.

The remaining two cases were neonatal infections in which lactobacilli may have been present but were not implicated in pathogenicity. Both infants responded to treatment (Dickgieser *et al.* 1984).

In conclusion, pathological conditions associated with lactobacilli are extremely rare and where they do exist, pre-existing disease are extremely rare and where they do exist, pre-existing disease states were present. Most cases were concurrent with oral pathology, lactobacilli being isolated probably as opportunists rather than primary pathogens.

## Plasmid Transfer Among the Lactobacilli

Despite the fact that lactobacilli have extrachromosomal genetic elements, these plasmids appear to be cryptic and stable.

### Plasmid Transfer

The streptococcal plasmid pAMß1 is a broad host range, promiscuous plasmid (Clewell *et al.* 1974) known, under laboratory conditions, to transfer among streptococcal groups A, B, D and H (Malke 1979), Bacillus spp (Landman *et al.* 1980; Orzech and Burke 1984). clostridia (Oultram and Young 1985), staphylococci (Engle *et al.* 1980) and certain lactobacilli (Gibson *et al.* 1979). Recent results (Morelli *et al.* 1988) have shown that transfer of pAMß1 can occur *in vivo*, however the animals used to demonstrate this transfer were germ-free mice associated only with the strains used in the experiment (*L. reuteri* and *Enterococcus faecalis*). Transformation rates were extremely low.

Shrago and Dobrogosz (1988) have reported the transfer of another streptococcal mutant plasmid pVA797 to *L. plantarum*. Attempts to use this plasmid to transform *L. plantarum*, however, failed since the plasmid vector underwent deletion in the *L. plantarum* cell. A plasmid has been isolated from *L. casei* and shown, under certain conditions, to transfer the ability to ferment lactose to other *L. casei* strains which had lost that ability (Tannock 1987).

It can be argued that if plasmid transfer between lactobacilli and other organisms in the gut occurred regularly *in vivo*, GI lactobacilli would be expected to harbour promiscuous plasmids such as pAMß1. Since this is not the case transfer of such plasmids either does not confer advantage or such transfer does not occur *in situ*.

In considering the use of live bacteria as a treatment for disease in humans certain factors must be taken into account. First, although the probiotic strain under consideration may have no history of pathogenicity, plasmid transfer is possible and antibiotic resistance or pathogenicity factors may be inherited from a freak event.

Secondly, lactobacilli have been found in association with certain pre-existing pathologies and this potential opportunism could lead to complications when treating patients rather than merely dosing normal individuals.

## Discussion

Results obtained here tend to suggest that pre-clinical screening of potential probiotics may be a useful way to ensure selection of organisms capable of surviving passage through conditions prevalent *in vivo*. Whether this applies to the ability of the organism to act clinically as a probiotic must await the result of properly conducted Phase 2 and Phase 3 studies. Comparison to probiotics which have similar antagonistic properties and with which clinical studies have been successfully conducted (Pozo-Olano *et al.* 1978; Vapaatalo *et al.* 1989; Vapaatalo *et al.* 1990) provides a powerful argument for the positive outcome of trials with *L. fermentum* KLD.

Neither study showed adverse reactions to *L. fermentum* KLD. The safety of dosing live lactobacilli appears to have no negative consequence for healthy volunteers. Literature, scant though it is, suggests that there is very little danger of pathogenicity due to lactobacilli under normal conditions or of the transfer of plasmids bearing pathogenicity or drug resistance factors being transferred to probiotic organisms *in vivo*. Some precautions might be required when dealing with patients. Pre-existing structural heart disease, severely carious teeth and pregnancy may be contra-indicated in the use of probiotics.

In the single dose volunteer study described here, three patterns of excretion of *L. fermentum* KLD were observed (Fig. 8.2). It is possible that these were manifestations of the design and small number of participants in the study. They may, however, be indicative of true excretion patterns for ingested bacteria. The volunteers excreting dosed organism during the first 48 h only may have particularly strong colonisation resistance, a large-intestinal milieu containing insufficient nutrients for the probiotic, an immune system particularly active against this type of organism or another factor predisposing short persistence of exogenous organisms in the gut. These factors may operate simultaneously and only against *L. fermentum* KLD or they may act against a wider range of non-commensal microbes. An individual eliminating *L. fermentum* in this way may retain another organism for a longer period of time.

At the other end of the scale was the volunteer who excreted *L. fermentum* KLD for 36 days. Why were conditions in this gut particularly suited to the persistence of this organism? Perhaps the individual harboured a bacterium very similar in some essential characteristic (other than plasmid profile) to *L. fermentum* KLD.

In the multiple dose study all volunteers had ceased to excrete *L. fermentum* KLD by 48 h after the last dose. It is only possible to speculate as to the reasons for this. The immune system would have had time to respond to the presence of an extraneous organism. Colonisation resistance factors may have built up with continued exposure to a non-commensal organism.

Whatever the reason, the results obtained are useful commercially. A probiotic which colonised the gut (rather than merely persisting for a short time) thus conferring life-long protection would not make commercial sense! Scientifically, however, one must question the clinical potential of a probiotic which was so poorly adapted to the actual conditions in the gut that it was unable to maintain a population without continuous dosage. These were healthy volunteers: perhaps in the gut of a patient suffering from diarrhoeal disease the probiotic may indeed provide protection from pathogens and act as a nucleus around which the normal gut microflora would quickly re-establish.

Some of the early damage done to the credibility of probiotics has been redressed by recent, properly conducted studies. However, with many questions still unanswered and many trials still failing to produce convincing clinical effects, new treatments such as antibody therapy may signal the end of the now lengthy probation period for probiotics.

# References

Allison C, McFarlan C, Macfarlane GT (1989) Studies on mixed populations of human intestinal bacteria grown in single and multistage continuous sulture systems. Appl Env Microbiol 55:679-683

Aronsson B, Mollby R, Nord CE (1985) Antimicrobial agents and *Clostridium difficile* in acute enteric disease. J Infect Dis 151:476-481

Bayer AS, Chow AW, Betts D, Guze LB (1978) Lactobacillemia - report of nine cases - important clinical and therapeutic considerations. Am J Med 64:808-813

Bruce AW, Reid G (1988) Intravaginal instillation of lactobacilli for prevention of recurrent urinary tract infections. Can J Microbiol 34:339-343

Chan RCY, Bruce AW, Reid G (1984) Adherence of cervical, vaginal and distal urethral normal microbial flora to human uroepithelial cells and the inhibition of adherence of gram-negative uropathogens by competitive exclusion. J Urology 131:596-601

Clements ML, Levine MM, Black RE, Robins-Brown RM, Cisneros LA, Drusano GL, Lanata CF, Saah AJ (1981) *Lactobacillus* prophylaxis for diarrhea due to enterotoxigenic *Escherichia coli*. Antimicrobial Agents and Chemotherapy 20:104-108

Clements ML, Levine MM, Ristaino PA, Daya VE, Hughes TP (1983) Exogenous lactobacilli fed to man - their fate and ability to prevent diarrheal disease. Prog Fd Nutr Sci 7:29-37

Clewell DB, Yagi Y, Duny GM, Schultz SK (1974) Characterisation of three plasmid DNA molecules in a strain of *Streptococcus faecalis*: identification of a plasmid determining erythromycin resistance. J Bacteriol 117:283-289

Cole CB, Fuller R, Carter SM (1989) Effect of probiotic supplements of *Lactobacillus acidophilus* and *Bifidobacterium adolescentis* 2204 on ß-glucosidase and ß-glucuromidase activity in the lower gut of rats associated with human faecal flora. Microbial Ecology in Health and Disease 2:223-225

Collins EB, Hardt P (1980) Inhibition of *Candida albicans* by *Lactobacillus acidophilus*. J Dairy Sci 63:830-832

Conway PL, Gorbach SL, Goldin BR (1987) Survival of lactic acid bacteria in the human stomach and adhesion to intestinal cells. J Dairy Sci 70:1-12

Cummings JH, Banwell JG, Englysst HN, Coleman N, Segal I, Bersohn D (1990) The amount and composition of human large bowel contents. Gastroenterol 98:A408

de Klerk HC, Smit JA (1967) Properties of a *Lactobacillus fermenti* bacteriocin. J Gen Microbiol 48:309-316

Dickgieser U, Weiss N, Fritsche D (1984) *Lactobacillus gasseri* as the cause of septic urinary infection. Infection 12:14-16

Engle HWB, Soedirman N, Rost JA, van Leeuwen WS, van Embden JDA (1980) Transferability of macrolide, lincomycin and streptogramin resistances between group A, B and D streptococci, *S. pneumoniae* and *Staph. aureus*. J Bacteriol 142:407-413

Foster TL, Winans L, Carski TR (1980) Evaluation of lactobacillus preparation on enterotoxigenic *E. coli*-induced rabbit ileal loop reactions. Am J Gastroenterol 73:238-243

Gibson Em, Chace NM, London SB, London J (1979) Transfer of plasmid-mediated antibiotic resistance from streptococci to lactobacilli. J Bacteriol 137:614-619

Gibson SAW, Macfarlane GT (1988) Characterisation of proteases formed by *Bacteroides fragilis*. J Gen Microbiol 134:2231-2240

Goldin BR, Gorbach SL, Saxelin M, Barakat S, Gualtieri L, Salminen S (1992) Survival of *Lactobacillus* species (strain GG) in human gastrointestinal tract. Dig Dis Sci 37:121-128

Goldin BR, Swenson L, Dwyer J, Sexton M, Gorbach SL (1980) Effect of diet and *Lactobacillus acidophilus* supplements on human fecal bacterial enzymes. J Nat Cancer Inst 64:255-261

Gombosova A, Demes P, Valent M (1986) Immunotherapeutic effect of the lactobacillus vaccine, Solco Trichovac, in trichomoniasis is not mediated by antibodies cross reacting with *Trichomonas vaginalis*. Genitourin Med 62:107-110

Gotz V, Romankiewicz JA, Moss J, Murray HW (1979) Prophylaxis against ampicilin-associated diarrhea with a lactobacillus preparation. Am J Hosp Pharm 36:754-757

Gotz VP, Romankiewics JA, Moss J, Murry HW (1979) Prophylaxis against ampicillin-induced diarrhoea with a *Lactobacillus* preparation. Am J Hosp Pharm 36:754-757

Grothe G, Koehler ME, Boedecker RH, Nietsch P (1989) Bioregulatory therapy of diarrhea: application drug monitoring of Bacillus subtilis spores. Therapiewoche 39:3300-3302

Heimdahl A, Nord CE (1979) Effects of phenoxymethylpenicillin and clindamycin on oral, throat and faecal microflora of man. Scand J Infect Dis 19:233-242

Horikawa Y (1986) Effects of *Lactobacillus casei*-containing ointment on the healing and protection against opportunistic infection of thermal injury wounds in mice. Hiroshima J Med Sci 35:1-14

Howe PR, Hatch RE (1917) A study of the microorganisms of dental caries. J Med Res 36:481-492

Jaspers DA, Massey LW, Luedecke LO (1984) Effect of consuming yogurts prepared with three culture strains on human serum lipoproteins. J Food Sci 49:1178-1181

Joerger MC, Klaenhammer TR (1986) Characterisation and purification of Helveticin J and evidence for a chromosomally determined bacteriocin produced by *Lactobacillus* helveticus 481. J Bacteriol 167:439-446

Klebanoff SJ, Belding ME (1974) Virucidal activity of $H_2O_2$generating bacteria: requirement for peroxidase and a halide. J Inf Dis 129:345-348

Landman OE, Bodkin DJ, Finn CW, Pepin RA (1980) Conjugal transfer of plasmid pAMß1 from *Streptococcus anginosis* to *Bacillus subtilis* and plasmid-mobilised transfer of chromosomal markers between *B. subtilis* strains. In: Polsinelli M and Mazza G (eds) Transformation. Cotswold Press Ltd, Oxford, England pp219-227

Loesch WJ, Straffon LH (1979) Longitudinal investigation of the role of *Streptococcus mutans* in human fissure decay. Infect Immun 26:498-507

Loguercio C, Del Vecchio Blanco C, Coltorti M (1987) Enterococcia strain SF68 and lactulose in hepatic encephalopathy: a controlled study. J Inter Med Res 15:335-343

Macfarlane GT, Cummings JH (1991) The colonic flora, fermentation, and large bowel digestive function. In: Phillis SF, Pemberton JH, Shorter RG (eds) The Large Intestine: Physiology, Pathophysiology and Disease. Raven Press, New York pp51-92

Malke H (1979) Conjugal transfer of plasmids determining resistance to marolides, lincosamides and streptogramin-B type antibiotics among group A, B, D and H streptococci. FEMS Microbiol Lett 5:335-338

Mann GV, Spoerry A (1974) Studies of a surfactant and cholesteremia in the maasai. Am J Clin Nutr 27:464-468

Mardh P-A, Soltesz LV (1983) *In vitro* interactions between lactobacillus and other microorganisms occurring in the vaginal flora. Scand J Infect Dis Suppl 40:47-51

McCormick EL, Savage DC (1983) Characterisation of *Lactobacillus* sp strain 100-37 from the murine gastrointestinal tract: Ecology, plasmid content and antagonistic activity towards Clostridium ramosum H1. Appl Env Microbiol 46:1103-1112

Michalek SM, Hirasawa M, Kiyono H, Ochiai K,McGhee JR (1981) Oral ecology and virulence of *Lactobacillus casei* and *Streptococcus mutans* in gnotobiotic rats. Infect Immun 33:690-696

Morelli L, Serra PG, Bottazzi V (1988) *In vivo* transfer of pAMß1 between *Streptococcus faecalis* and *Lactobacillus reuteri*. J Appl Bact 65:371-375

Nagy E, Petterson M, Mårdh P-A (1991) Antibiosis between bacteria isolated from the vagina of women with and without signs of bacterial vaginosis. APMIS 99:739-744

Naudé W du T, Swanepoel A, Bohmer RH, Bolding E (1988) Endocarditis caused by *Lactobacillus casei* subspecies *rhamnosus*. S Afr Med J 73:612-614

Nord CE, Edlund C (1991) Ecological effects of antimicrobial agents on the human intestinal microflora. Microbial Ecology in Health and Disease 4:1933-207

Nord CE, Kager L, Heimdahl A (1984) Impact of antimicrobial agents on the gastrointestinal microflora and the risk of infections. Am J Med 76:99-106

Orzech KA, Burke WF (1984) Conjugal transfer of pAMß1 in *Bacillus sphaericus* 1593. FEMS Microbiol Lett 25:91-95

Oultram JD, Young M (1985) conjugal transfer of pAMß1 from *Streptococcus lactis* and *Bacillus subtilis* to *Clostridium acetobutylicum*. FEMS Microbiol Lett 27:129-134

Pattersson L, Graf W, Sewelin U (1983) Survival of *Lactobacillus acidophilus* NCDO 1748 in the human gastrointestinal tract. In: XV Symposium Swedish Nutrition Foundation. Hallergren B (ed) Almquist and Wilssell, Uppsala, Sweden pp127-130

Perrons CJ, Donoghue HD (1990) Colonisation resistance of defined bacterial plaques to *Streptococcus mutans* implantation on teeth in a model mouth. J Dent Res 69:483-488

Pozo-Olano JD, Wanan JH, Gomez RG, Cavazos MG (1978) Effect of a lactobacilli preparation on traveller's diarrhea. Gastroenterology 74:829-830

Raccach M, McGrath R, Daftarian H (1989) Antibiosis of some lactic acid bacteria including *Lactobacillus acidophilus* toward *Listeria monocytogenes*. Int J Food Microbiol 9:25-32

Reuman PD, Duckworth DH, Smith KL, Kagan R, Bucciarelli RL, Aijoub EM (1986) Lack of effect of *Lactobacillus* on gastrointestinal bacterial colonisation in premature infants. Pediatr Infect Dis 5:663-668

Salminen E, Elomaa I, Minkkinen J, Vapaatalo H, Salminen S (1988) Preservation of intestinal integrity during radiotherapy using live *Lactobacillus acidophilus* cultures. Clinical Radiother 39:435-437

Sandler B (1979) Lactobacillus for vulvovaginitis. The Lancet Oct 13:791-792

Saxelin M, Elo S, Salminen S, Vapaatalo H (1991) Dose response colonisation of faeces after oral administration of *Lactobacillus casei* strain GG. Microbial Ecology in Health and Disease 4:209-214

Shrago AW, Dobrogosz WJ (1988) Conjugal transfer of Group G streptococcal plasmids and comobilisation of *E. coli* - Streptococcus shuttle plasmids to *L. plantarum*. Appl Env Microbiol 54:824-826

Silva M, Jacobus NV, Deneke C, Gorbach SL (1987) Anitmicrobial substance from a human lactobacillus strain. Antimicrob Agents Chemother 31:1231-1233

Smith HW (1975) Survival of orally administered *E. Coli* K-12 in the alimentary tract of man. Nature 255:500-504

Steffen R, Heusser R, Du Pont HL (1986) Prevention of travellers' diarrhea by nonantibiotic drugs. Rev Inf Dis 8:5151-5159

Tannock GW (1987) Conjugal transfer of pAMß1 in *L. reuteri* and between lactobacilli and *Enterococcus faecalis*. Appl Env Microbiol 53:2693-2695

Tannock GW, Fuller R, Smith SL, Hall MA (1990) Plamid profiling of members of the family *Enterobacteriaceae*, lactobacilli, and bifidobacteria to study the transmission of bacteria from mother to infant. J Clin Microbiol 28:1225-1228

Tompson LU, Jenkins DJ, Amer MV, Reichert R, Jenkins A, Kamulsky J (1982) The effect of fermented and unfermented milks on serum cholesterol. Am J Clin Nutr 36:1106-1108

Tramer J (1966) Inhibitory effect of *Lactobacillus acidophilus*. Nature 211:204-205

Turnbull PCB, Gerson PJ, Stanley G (1978) Inability of selected lactobacilli to inhibit heat-labile or heat-stable enterotoxin effect of *Escherichia coli* B7A. J Appl Bact 45:157-160

Upreti GC, Hindsill RD (1973) Isolation and characterisation of a bacteriocin from a homofermentative *Lactobacillus*. Antimicrob Agents Chemother 4:487-494

Upreti GC, Hindsill RD (1975) Production and mode of action of lactocin 27:bacteriocin from a homofermentative *Lactobacillus*. Antimicrob Agents Chemother 7:139-145

Vapaatalo H, Salminen S, Isolauri E, Saxelin S, Gorbach S (1990) The use of a human Lactobacillus strain (Lactobacillus GG) in the prevention and treatment of diarrhoea. XV International Congress on Microbial Ecology and Disease pp125

Vapaatalo H, Salminen S, Siitonen S, Gorbach S (1989) *Lactobacillus* GG in the prevention of antibiotic associated diarrhea. IV World Conference on Clinical Pharmacology and Therapeutics pp432

Will TE (1979) Lactobacillus overgrowth for treatment of moniliary vulvovaginitis. The Lancet Sept 1:482

Wilson KH, Perini F (1988) Role of competition for nutrients in suppression of *Clostridium difficile* by the colonic microflora. Infect Immun 56:2610-2614

Yost RL, Gotz VP (1985) Effect of a *Lactobacillus* preparation on the absorption of oral ampicillin. Antimicrob Agents Chemother 28:727-729

Chapter 9

# The Therapeutic use of Live Bacteria in Newborn Infants

M. A. Hall and S. L. Smith

## Introduction

As we move inexorably towards the 21st century and its attendant technological triumphs the paediatrician of iconoclastic persuasion could reasonably be forgiven for being tempted to question the assertion that breast feeding provides the optimal form of nutrition for infants in the developed world. For, despite the undeniable benefits of breast feeding which are vital to the survival of babies in the developing world, there is evidence that, in less impoverished countries, babies fed on modified cow's milk formulae gain more weight than their breast-fed counterparts (Peerson *et al.* 1992), that they are at reduced risk of developing conditions such as "haemorrhagic disease of the newborn" (McNinch and Tripp 1991) and iron deficiency (Calvo *et al.* 1992) and that their sleeping patterns may be more regular than those who are breast-fed (Elias *et al.* 1986).

A paediatrician propounding such a view in the nineteenth century would not only have been regarded as iconoclastic, he could justifiably have been accused of frank malpractice. For, prior to the introduction of modified cow's milk for infant feeding, just before the second world war, those babies who were unable to be fed with human milk were at serious risk of dying as a result of malnutrition or infection. It is not surprising, therefore, that Tissier's discovery of a characteristic type of microbial flora in the faeces of breast-feeding infants (Tissier 1900) led many to believe that, if the bowel flora of babies who were fed with cow's milk could be changed to approximate that of breast-feeding infants, there would be an associated improvement in their general health and chances of survival. Since that time there have been many attempts to induce a modification of the intestinal flora of human infants both by adaptation of the components of artificial formulae and by the administration of probiotics.

More recently, with the impressive advances which have occurred in the medical care of premature babies, there has been something of a revival of interest in the potential role of probiotics for the modification of bacterial colonisation at various anatomical sites, including the skin, the upper respiratory tract and the intestine. The purpose of this chapter is to review some of the work which has been reported concerning the administration of probiotics to human infants in the newborn period.

## Studies of Full-term Infants

### Lactobacilli

One of the central hypotheses propounded by Elie Metchnikoff, then sub-director of the Pasteur Institute, in his much-quoted book of 1907 (Metchnikoff 1907), was that the cause of "precocious and unhappy old age" was the presence in the intestine of "microbes of putrefaction" which he believed released toxins capable of accelerating the ageing process. It was Metchnikoff's belief that such intestinal putrefaction and, therefore, the progression of the ageing process, could be arrested by the ingestion of agents which reduced the growth of the putrefying intestinal microbes; he recommended "the absorption either of soured milk prepared by a group of lactic bacteria, or of pure cultures of the Bulgarian bacillus, but in each case taking at the same time a certain quantity of milk sugar or saccharose", a diet which he himself had taken with some success for several years. Such advice was not based on mere theorising; a considerable body of evidence had already been amassed which indeed appeared to demonstrate a reduction in "microbes of putrefaction" following either the ingestion (Cohendy 1906) or, in experiments with dogs (Herter 1897), the injection into the intestine, of lactic acid bacteria. There was also some early evidence that certain potentially fatal infections, such as cholera and tuberculosis, could be prevented by the administration of lactobacilli (Metchnikoff 1894).

The interaction of lactic acid bacteria with coliform organisms was the subject of numerous studies during the first half of the 20th century and there were many reports of reciprocal relationships between these organisms. In particular, studies involving pigs indicated that the administration of lactic acid bacteria was associated not only with a suppression of *E. coli* in the bowel flora but also with an improvement in weight gain (Mollgaard 1946). These findings were supported by later studies (Kershaw *et al.* 1966, Fuller 1989). It is not surprising, therefore, that the earliest recorded studies of bacterial therapy in newborn infants involved the administration of lactobacilli, since this organism was deemed to be non- pathogenic. One of the first scientifically conducted studies of the effects of probiotics fed to newborn infants took place in Ohio soon after the second World War, at a time when there was no doubt that breast-fed infants thrived better than those fed formula of the same protein level (Robinson and Thompson 1952). In what appears to have been a randomised and controlled study of newborn infants Robinson and colleagues recruited all infants delivered over a six month period in two different hospitals. These infants were divided into eight groups: Group 1 (n=123) were the controls and received no addition to the daily formula; Group 2 (n=129) received formula with folic acid added (2mg per quart of milk); Group 3 (n=124) received formula with a mixture of *Lactobacillus acidophilus* and *Lactobacillus bifidus* added in a dose of $5 \times 10^8$ organisms per quart of milk; Group 4 (n=134) received formula with both folic acid and the lactobacillus mixture added. The supplementations to the formula were made only during infants' stay in hospital - usually between one and six days. The infants in groups 5-8 were all partially breast fed during the first month of life. Group 5 (n=69) were the partially breast-fed controls; Group 6 (n=83) were partially breast-fed with additional folic acid; Group 7 (n=79) were partially breast-fed with *Lactobacillus bifidus* added in the same quantity as the formula fed infants and Group 8 (n=60) received both folic acid and the lactobacillus preparation. It was found that supplementation of infants fed on only artificial milk during the first one to six days of life with the mixture of *Lactobacillus acidophilus* and *Lactobacillus bifidus* was associated with a greater gain in weight than in formula-fed babies who did not receive such supplementation; no effect of supplementation was found in babies who were partially breast-fed. This study also confirmed that weight gain was significantly better in babies who received breast milk than in those who were formula-fed, even if it was only for one or two days. However,the findings of this study are not easy to interpret since there was no attempt at correlation between the clinical findings and the bacterial content of the stool; no information was provided concerning the formulation of the artificial milks which were used and information concerning weight gain during the first month was available for only about 800 of the 1400 or so babies enrolled into the study.

Since this report there have been several published studies of the effect of orally administered lactobacilli and bifidobacteria to infants with diarrhoea (Tomic-Karovic and Fanjek 1962, Aritaki and Ishikawa 1962, Pearce and Hamilton 1974, Hotto *et al.* 1987, Isolauri *et al.* 1991), the results of which have been variable. To date, however, there is no convincing evidence that the administration of such organisms leads to a measurable long-term improvement in the well-being of normal babies or is associated with a consistent pattern of bacterial colonisation in the lower intestine.

## Staphylococcus aureus

There can be little doubt that an important factor leading to the reduction in perinatal mortality which has been seen in many developed countries in the 20th century was the introduction of the policy of delivering almost all babies in a hospital or appropriately equipped maternity unit. However, one of the unwanted spin-offs of this policy, which was inevitably associated with babies being nursed in close proximity to each other and handled by staff colonised with potentially pathogenic bacteria, was the emergence of epidemics of serious nosocomial infection. One of the main early bacterial causes of such sepsis was *Staphylococcus aureus*. Infection with this organism could cause conditions ranging from relatively minor skin sepsis to fatal blood poisoning (septicaemia) or meningitis and in the 1950s and 1960s the frequency of such infections reached epidemic proportions (Gillespie *et al.* 1958). Initially these infections were readily treatable with penicillin but with the widespread use of the beta-lactam antibiotics organisms emerged which were capable of producing beta-lactamase enzymes which rendered the bacteria resistant to antibiotics such as penicillin. Although beta-lactamase resistant antibiotics were developed there was concern that bacteria such as staphylococci would eventually become resistant to all available antibiotics and alternative strategies for reducing the risk of newborn babies becoming colonised with pathogenic staphylococci were tried.

One such strategy - the administration of an apparently harmless strain of *Staphylococcus aureus* - represented one of the first attempts at probiotic administration to at-risk newborn babies based on the concept of "bacterial interference". This approach was devised as a result of epidemiological surveillance among babies and staff in a nursery in the New York Hospital in 1959 (Shinefield *et al.* 1963). It was found that one of the nurses on the unit was a nasal carrier of *Staphylococcus aureus* phage type 80/81 and that 22% of babies who were handled by the nurse within 24 h of birth became colonised with the same strain; in comparison, no baby who first came into contact with this nurse after the age of 24 h became colonised with phage type 80/81 although 84% of these babies were found to be colonised with other strains of *Staphylococcus aureus*. It was inferred, from these findings, that the presence of staphylococci at a single site interfered with subsequent acquisition of other strains of staphylococci at that particular site and that, if a non-pathogenic strain of staphylococcus could be administered, the process of "bacterial interference" could be used to prevent colonisation with pathogenic strains of staphylococci (Shinefield *et al.* 1963).

The non-pathogenic strain of *Staphylococcus aureus* - subsequently labelled strain 502A - developed by these investigators also originated from the nasal mucosa of a nurse working on the nursery. More than 100 babies were identified as having become colonised with 502A as a result of contact with the nurse and these babies were monitored for a period of more than one year, during which time no evidence of staphylococcal sepsis was recorded in either the babies or their families. Strain 502A was, therefore, considered to be "safe" and further studies were undertaken which indicated that the "dose" required to effect colonisation of the umbilicus in 50% of newborn infants was 10 cells or less and 250 cells for the nasal mucosa (Shinefield *et al.* 1963). Randomised controlled trials demonstrated that such artificial colonisation was effective in protection against staphylococcal sepsis with phage type 80/81 during the first year of life and this strategy was subsequently adopted to control epidemics of staphylococcal infections in newborn nurseries (Shinefield *et al.* 1971, Light *et al.* 1965, Light *et al.* 1967).

In any probiotic study involving human subjects two of the main safety criteria must be that there should be no detrimental effect on the microbial ecology of the human host and that the probiotic organism should not itself be pathogenic. For 502A it was not long before it became clear that it was not as safe as it had at first seemed. First, it was observed that up to 34% of infants who were artificially colonised with 502A developed small vesicles around the umbilicus, although these did not appear to cause any discomfort or change in the babies and they spontaneously disappeared (Albert *et al.* 1970); other infants developed conjunctivitis which was usually resolved within a few days. However, more seriously, at least one case of fatal septicaemia and meningitis occurred in a baby who had been treated with strain 502A (Houck *et al.* 1972). With the development of alternative strategies, such as the judicious use of umbilical antiseptic agents and appropriate cross-infection control measures, the incidence of staphylococcal infections declined from the mid 1970s and the need for "bacterial interference" programmes disappeared (Goldmann 1988).

## Studies of "High-Risk" Newborn Babies

The past 25 years has seen a dramatic change in the chances of survival of babies who are born prematurely as well as certain full-term babies who develop problems such as bacterial infection. One of the main factors which has enabled these improvements to occur has been the development of intensive care, and neonatal intensive care units, for sick newborn infants. Thus, with the increasing survival of extremely premature babies, there has emerged, in effect, a new population of infants who are initially nursed in an environment which is not only quite different to that in which normal full-term infants are nursed, but which also may harbour bacterial pathogens which are found almost exclusively in the hospital setting.

## Ecological Implications of Neonatal Intensive Care

One of the clinical features which distinguishes newborn babies from older patients is their susceptibility to certain types of bacterial infection. Not only is there an increased risk of becoming infected but babies may become moribund within an hour or two of the first signs of infection when organisms such as the group B Streptococcus are involved. Although the incidence of proven bacterial sepsis is actually low, the clinical signs of infection are often indistinguishable from other more common types of illness, such as the respiratory distress syndrome. For these reasons, despite the low incidence of true bacterial infections, broad- spectrum antibiotic treatment is usually started as soon as a newborn baby develops significant signs of being unwell. Consequently, antibiotic usage is widespread in neonatal units (Hall 1990) and, inevitably has a profound influence on the bacterial flora, not only of the individual babies, but also of the immediate nursery environment. In addition, there are many other aspects of neonatal intensive care which influence the bacterial flora of babies receiving intensive care, including incubator nursing (Hall *et al.* 1990) and the topical application of anti- septic agents such as hexachlorophene to the umbilical cord (Gillespie *et al.* 1958).

As a result of these interventions, the neonatal intensive care nursery is an environment which harbours bacteria such as *Staphylococcus epidermidis* and coliform organisms which are not only potential pathogens for these vulnerable babies but also may have become resistant to the antibiotics which are used to treat suspected sepsis in such babies. There is good evidence, too, that such organisms contribute to the bowel flora and that the stool bacteria of preterm babies and others receiving intensive care differs significantly from that of normal full-term babies (Hall *et al.* 1990, Bennet *et al.* 1986, Sakata *et al.* 1985). Penetration of these bacteria to sites such as the lungs and the blood is facilitated by some of the invasive procedures which are used in the management of neonates receiving intensive care, such as the use of tubes to connect the babies to artificial ventilators (endotracheal tubes) and of intravenous catheters which are placed near to, or within, the heart. It is possible, therefore, that active modification of the environmental bacterial flora to one which is benign, rather than pathogenic, could reduce the risk of serious sepsis in newborn babies; it is also possible that the bowel flora may have important physiological functions which have yet to be defined and that appropriate modifications may have a therapeutic role in the management of sick newborn infants.

## Nasopharyngeal Implantation of Alpha Haemolytic Streptococcus

The first serious contemporary attempt to apply bacterial therapy to "at-risk" newborn infants was reported by Dr. Katherine Sprunt and her colleagues working at the Babies Hospital in New York. Their findings from surveillance studies of the oropharyngeal bacterial flora of 223 neonates in the Neonatal Intensive Care Unit (NICU) indicated that clinical infection occurred in babies who showed an abnormal pattern of bacterial colonization of the pharynx; in these babies, organisms with the same cultural characteristics as those causing the infection could be found in the oropharynx (Sprunt *et al.* 1980). In contrast, the majority of infants whose oropharyngeal cultures grew alpha haemolytic streptococci did not become infected; the exceptions were two babies who became colonised with, and infected by, *Staphylococcus aureus*. This led these workers to define "normal" oropharyngeal bacterial colonization as the presence of alpha haemolytic streptococci as predominant organisms in quantities exceeding $10^4$ cfu/ml of a one-ml culture sample. "Abnormal" colonization (or bacterial "overgrowth") was defined as the presence of a predominant organism, other than alpha haemolytic streptococci, in a concentration $10^4$ or greater and constituting >90% of the bacterial population; for *S. aureus* this definition was modified, and "abnormal" colonization was defined as the presence of $10^3$ cfu/ml or more which were not necessarily the predominant organism but which constituted >2% of the total bacterial population. An inevitable inference of these findings was that, if a "normal" oropharyngeal flora could be induced, the risk of bacterial infection in neonates residing in the NICU may be reduced. Dr. Sprunt and her colleagues, therefore, decided to undertake a study involving the nasopharyngeal implantation of a strain of alpha-haemolytic streptococcus in neonates at risk of developing bacterial infection in order to induce a "normal" bacterial flora (Sprunt et al. 1978). The criteria laid down for the implant strain required that the organism should be:

i obtained from the oropharynx of normal newborn infants who remained healthy for at least three months after birth;
ii resistant to 0.5 to 1.0 microgram, but not 3.0 microgram of ampicillin per ml and sensitive to a number of antibiotics;
iii able to inhibit *in vitro* growth of organisms causing abnormal colonisation, such as *S. aureus*, enteric type gram-negative bacilli and Pseudomonas aeruginosa;
iv consistent with assignment to the mitis group of viridans streptococci.

After extensive testing a strain was found which fulfilled these criteria - strain no 215 - and solutions containing $10^6$ cfu/0.1 ml were prepared. The organisms were implanted, in a dose of 2 x $10^5$ to 5 x $10^6$ cfu, into the nasopharynx of 22 neonates whose weight on the day of implant ranged from 700 grams to 2850 grams and whose ages ranged from eight to 64 days. All of these infants were considered to be at high risk of infection because they were found to harbour potential pathogens - including gram-negative organisms and staphylococci - in the nasopharynx. In 16 of the 22 infants streptococci were recovered from the nasopharynx for a variable period following implantation although testing of the identifying marker characteristics revealed that it was not always the implant strain which was present; it appears, therefore, that the administration of the implantation strain acted as a "recruiting agent", serving to encourage colonisation with alpha haemolytic streptococci of varying strains. In this small number of babies, no adverse effects were recorded but further studies of the effects of implantation of alpha haemolytic streptococci were not undertaken by this group, apparently because of a significant reduction in the incidence of gram-negative infections, the previous high level of which had been the motivation for undertaking these studies (Goldmann 1988).

Similar studies were performed by Cook *et al.* in the Louisville General Hospital who reported that an outbreak of gram-negative infection had occurred in the intensive care nursery over a period of three months. Eleven neonates were colonised or infected with either *Klebsiella spp.*, *Enterobacter spp.*, or *Serratia spp.* (KES). The organisms were resistant to most antibiotics, including amikacin, sissomicin, and netilimicin. Of these 11 neonates, three in the intensive care room had pharyngeal implantation of the alpha haemolytic streptococcus. These three infants showed no growth of this organism in their nasopharynx prior to inoculation. In addition, an infant in the "clean room" received implantation of alpha haemolytic streptococcus. The purpose of this inoculation was to initiate growth of a normal range of bacterial flora and so decrease the risk of acquiring resistant KES colonisation and subsequent infection. In all infants, overgrowth of alpha streptococcus and elimination of KES from the nasopharynx was achieved. Of the eleven infants initially infected with a bacterium of the KES group, three died; two of these babies had received implantation: in one KES had been successfully eradicated but the infant died as a result of chronic lung disease and intracranial haemorrhage; in the second eradication of KES was initially successful but colonisation with *Klebsiella spp.* and *Serratia spp.* recurred and the baby died as a result of *Serratia marcescens* septicaemia and meningitis.

## Studies of Intestinal Implantation

There is general agreement that one of the main reservoirs of potentially pathogenic bacteria for "at-risk" newborn infants is their own lower intestine. Organisms which are not only pathogenic but which also may be resistant to a wide range of antibiotics may rapidly become established within the bowel of these infants. It is possible that one way of preventing such colonization is to introduce into the digestive system a benign bacterial strain, which is capable of inhibiting the proliferation of pathogenic bacteria. In recent years two types of organism have been used to investigate this possibility - *E. coli* and lactobacilli.

### *E. coli*

Although E.coli is a well-recognised cause of sepsis in newborn infants, certain strains of this organism have been used to prevent colonization of the bowel by pathogenic bacteria. A group of researchers in Tours, France, have identified strains of *E. coli* (strains"ECA" and "EMO") which have no K capsular antigens and which have been found, in animal experiments, to be non-pathogenic; further, these strains have been shown to be capable of colonizing the bowel of healthy full-term infants (Borderon *et al.* 1978, Duval-Iflah *et al.* 1983). Preliminary studies have indicated that colonisation of the bowel of newborn infants with these organisms is more likely to be successful if the bacteria are administered soon after birth, particularly within the first two hours (Poisson *et al.* 1986, Duval-Iflah *et al.* 1982). In an open study Rastegar Lari *et al.* (1990) investigated the effect of administering these strains of *E. coli* to preterm infants whose gestational ages ranged from 31-37 weeks and birthweights 1.4kg - 2.4kg. Infants were randomly assigned to one of four groups: group 1 (16 infants) received *E. coli* ECA; group 2 (16 infants) received *E. coli* EMO; group 3 (16 infants) received both ECA and EMO and group 4, the control group, received neither organism. The study bacteria were administered as a suspension of $10^5$ organisms in 1 ml of water, mixed with milk; two doses were given - the first within 12 h of birth and the second 12 h later. None of these babies received treatment with antibiotics during the study period. The results of this study are represented in Figures.9.1a and 9.1b which show

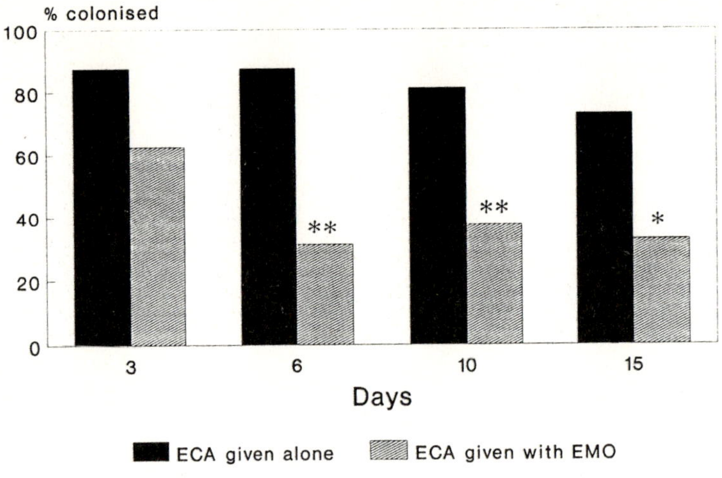


**Fig.9.1a.** Implantation and persistance of administered strains at a level ≥ $10^7$/g of faeces: % of newborns colonised. ■ single administration of ECA; □ ECA and EMO. Significant differences: *p 0.05-0.01; **p<0.01

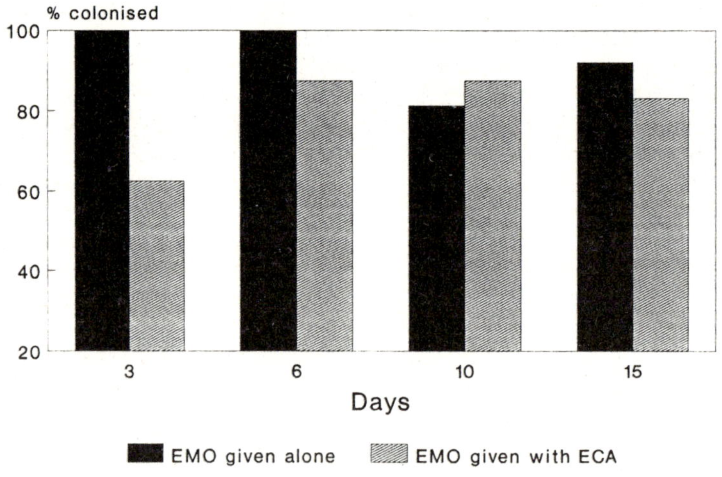


**Fig. 9.1b.** Implantation and persistance of administered strains at a level ≥ $10^7$/g of faeces: % of newborns colonised. ■ single administration of EMO; □ EMO and ECA.

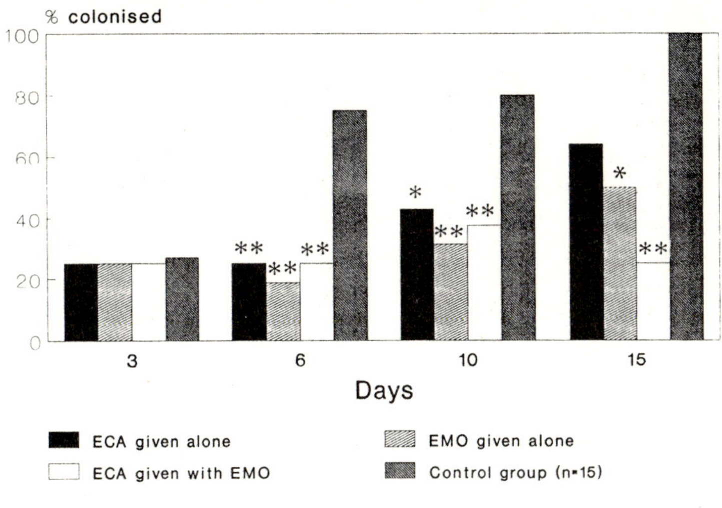


**Fig. 9.2.** Comparison between the four groups: % of newborns colonised with ≥ ore resistant enteric organism at a level $\geq 10^7$/g of faeces. Significant differences: *p0.05-0.01; **p<0.01.

the colonisation rates with the study organisms in the three groups, where "colonisation" is defined as the presence of at least $10^7$ organisms per gram of faeces; it can be seen that the majority of babies became colonised with the study organism when administered alone and that colonisation lasted for at least 15 days in 73% of those receiving ECA and 92% of those who received EMO. When both were given, the colonisation rates at 15 days were 85% for EMO and 33% for ECA.

In Figure 9.2 the rates of colonisation with antibiotic-resistant organisms in the three groups are shown; it can be seen that infants receiving EMO, either singly or in combination with ECA, were less likely than those in the control group to be colonised with antibiotic-resistant organisms between the ages of six days and 15 days while this effect was present on days 6 and 10, but not day 15, for those receiving ECA alone.

The results of this study indicate that, although both ECA and EMO are capable of colonising the bowel of premature infants, EMO was more likely to persist for longer than ECA; further, EMO appears to have a suppressive effect on the growth of ECA in this group of babies. The reason for the apparent dominance of ECA is not clear but it may be of relevance that the EMO strain is a lactose-fermenter. None of the babies in this study were treated with systemic antibiotics, however, and it is likely, therefore, that they were not representative of the sickest of babies receiving neonatal intensive care. There is a need for further studies in newborn populations before the safety and potential role of these organisms can be established.

## *Lactobacilli*

The concept of administering lactobacilli to babies has been discussed over many years but there have been few reported scientific studies in the English language literature of the effects of the administration of such organisms to preterm babies. However, this has not been the case in Europe where in 1964 a report was published by Vicek and Kneifl (Vicek and Kneifl 1964). In this study capsules of "Omniflora" were administered twice daily to preterm infants. Omniflora consisted of a commercial mixed culture of " *Lactobacillus acidophilus*, *A. bifidus* and a non-pathogenic *E. coli*. These organisms became established in the intestines of 20 out of 24 preterm infants within one week.

More recently a study has been reported by Reuman *et al.* from a neonatal unit in Gainesville, Florida. In this study preterm babies whose gestational ages ranged from 25 weeks - 34 weeks and who weighed less than 2000 grams at birth were included. The organism under investigation, *Lactobacillus acidophilus*, was selected because it had been used safely in previous studies involving human subjects and because a commercial preparation was available. This study was a double-blind placebo-controlled trial in which the study infants were randomised to receive either *Lactobacillus acidophilus*- containing formula or non-lactobacilli containing formula from 72h of life. The lactobacilli were prepared in 1-mL vials of milk formula, containing approximately $10^8$ organisms, and administered orally twice daily for a period of time which is unspecified but which is likely to have been for the duration of the infants' stay on the unit; the control group received 1-ml vials of formula which did not contain lactobacilli. The two study groups were screened weekly, by means of rectal swabs, for stool lactobacilli and gram- negative bacteria; antibiotic resistance in the stool organisms was also determined. Of the 15 babies who received lactobacilli, 13 were found to have lactobacilli in the stool, compared with only three of the 15 controls; although this difference was statistically significant ($p<0.001$), there was no significant difference between the groups in the frequency of isolation of gram-negative organisms from the stool or in the frequency of antibiotic resistance in these organisms. Further, there was no difference between the groups in the clinical course of the babies or in weight gain. The lactobacilli first appeared in the stool at a mean age of 19 days in the treated group and 47 days in the three control babies; all babies in both groups were treated systemically with gentamicin and ampicillin for at least one day. It is also of note that only two infants in the test group and one in the control group received breast milk. No quantitive stool bacteria was reported from this study and no attempts were made to determine whether the lactobacilli found in the stool were the same strain as those administered to the babies.

In Southampton the relevance of the bowel bacterial flora to the well-being of preterm infants has been a subject of developing interest over the past few years. In a preliminary study of faecal bacteria, we found that, in comparison with healthy full-term infants, preterm babies were less likely to show evidence of intestinal colonisation with lactobacilli, particularly if they were nursed in incubators or if they had received broad-spectrum antibiotics; this was in contrast to coliform organisms which were found to be present in relatively large numbers in both the full-term and the preterm populations (Hall *et al.* 1990). The significance of these findings is not clear but, although there is no firm evidence that this lack of lactobacilli in preterm infants is detrimental, it is possible that intestinal lactobacilli may have a role in nutrient utilisation or control of bowel flora which is physiologically beneficial to such babies.

In order to explore this further we decided to investigate the possibility of administering lactobacilli to preterm infants who required admission to the neonatal intensive care unit. After some deliberation we elected to use *Lactobacillus GG* as the test organism. This organism was selected for three reasons: first, it has been demonstrated to fulfil the criteria required for bowel colonisation (Silva *et al.* 1987, Goldin *et al.* 1992); second, although *Lactobacillus GG* has not been reported to have been given to newborn infants, it has been used to treat relatively young infants suffering from acute gastroenteritis (Isolauri *et al.* 1991) with no reported adverse effects; finally, *Lactobacillus GG* is readily identifiable from stool cultures and it can be inferred, therefore, that the presence of the organism in the stool has resulted from the oral administration. The initial phase of our studies was aimed at determining:

i whether *Lactobacillus GG*, administered orally to preterm newborn infants, results in recovery of the organism from the stool;
ii the dose required to establish stool colonization in significant quantities;
iii whether any recognisable adverse effects occur following the oral administration of *Lactobacillus GG* to preterm infants.

For this initial phase three infants were selected whose gestational ages ranged from 25 to 35 weeks with an average of 33 weeks. The birth weight range was 860 gms - 2130 gms with a mean weight of 1460 grams. This initial part of the study was not blinded in order to allow the clinicians the opportunity to evaluate fully any changes in the infants's condition.

*Lactobacillus GG* was supplied as a freeze-dried powder containing $10^{11}$ cfu/g dry weight. Every day the solution was prepared by suspension of the freeze-dried powder in sterile distilled water. This dilution was then added to the infant's milk feed twice daily. The first dose of the organism was usually given with the initiation of milk feeds. Samples of the supplemented milk were sent for bacterial analysis daily to ensure that there was no contamination with other bacteria. Daily faecal samples were collected and quantitative bacteriology was performed on these specimens. Identification of isolates was performed using standard laboratory methods and isolates initially described as *Lactobacillus GG* were further speciated and defined to confirm that the organism given orally was the same organism cultured from the stool specimens. In addition, urine specimens were analyzed at least weekly for the presence of lactobacilli. A daily clinical assessment was also performed. This involved not only a physical examination but clinical information regarding stool frequency, girth measurements, any milk intolerance, vomiting or other abnormal findings was recorded.

In this "dose-ranging" study the first infant studied received $10^4$cfu/ml *Lactobacillus GG* for five days; this resulted in convincing stool colonisation with counts ranging from $10^4$ cfu/g dry weight of stool in the first week following administration to $10^6$ cfu/g dry weight of stool five weeks later with a maximum count $10^8$ cfu/g dry weight of stool two weeks after receiving the first dose of *Lactobacillus GG*; the next infant received $10^6$ cfu/ml organisms for five days and in this infant colonisation was not so consistent - indeed *Lactobacillus GG* was isolated on only one occasion at a count of $10^{10}$ cfu/g dry weight of stool. The final infant received $10^6$ cfu/ml twice daily for 14 days. *Lactobacillus GG* was isolated again on only one occasion at a count of $10^9$ cfu/g dry weight of stool. All of these infants received a variety of milk and no infant was exclusively breast-fed from birth. In addition all three infants were treated with antibiotics.

In this small study there appeared to be no significant beneficial or detrimental clinical effects, with regard to intensive care requirements, weight gain, frequency or consistency of stools, milk intolerance or discharge home. The study did, however, confirm that when *Lactobacillus GG* is administered orally stool colonisation with the organism can be achieved, but this may be variable and may not be in significant quantities. Subsequent to these preliminary studies we conducted a randomised, double-blind study of the effects of *Lactobacillus GG* in preterm infants who were given a dose of $10^8$ cfu/ml twice daily for 14 days. The results of this study are now in the process of being analyzed.

## Conclusion

The concept of modifying the bacterial flora of the bowel of newborn infants has been a subject of considerable interest and research for well over a century. The observation that breast-feeding babies have a stool bacterial flora which is different to that of "bottle-fed" babies has inevitably led to speculation that modification of the stool flora may, in some way, benefit those babies who are not able to breast feed. While *in vitro* and animal experiments have often been successful in demonstrating that it is possible to reduce colonization with coliform bacteria by introducing organisms such as lactobacilli, the published work involving human infants has been less convincing. Evidence that in the human infant, administration of lactic acid bacteria is associated with a measurable clinical or physiological benefit is, to date, almost completely lacking.

The neonatal intensive care unit is something of a microbiological minefield with the potential for the development and rapid spread of antibiotic-resistant pathogens. The use of "benign" bacteria to effect a biological reduction in such organisms is not without serious risk, as illustrated by the administration of *Staphylococcus aureus* - strain 502A. Even lactobacilli - generally described as a benign organism - can be pathogenic. However, the investigations described above indicate that the principle of bacterial interference can be applied to reducing colonisation with pathogenic organisms at three sites: the umbilicus and skin, using *Staphylococcus aureus* - strain 502A; the upper respiratory tract, using Streptococcus viridans; the lower bowel, using *E. coli*, strains ECA and EMO. To date however the use of chemical agents, in the form of antibiotics and skin disinfectants, has obviated the need for the widespread introduction of therapies involving the administration of live bacteria - therapies perceived to carry risks to both the individual patient and the ecological environment. A role for the use of live bacteria in the routine management of the newborn has yet, therefore, to be defined.

## References

Albert S, Baldwin R, Czekajewski S et al (1970) Bullous impetigo due to group II *Staphylococcus aureus*. Amer J Dis Child 120:10-13

Aritaki S, Ishikawa S (1962) Application of lactobacillus acidophilus fermented milk in the pediatric field. Acta Pediatr Jap 66:811-815

Bennet R, Eriksson M, Nord CE, Zetterstrom R (1986) Fecal bacterial microflora of newborn infants during intensive care management and treatment with five antibiotic regimens. Pediatr Infect Dis 5:533-9

Blair EB, Tull AH (1969) Multiple infections among newborns resulting from colonisation with *Staphylococcus aureus* 502A. Amer J Clin Path 52:42-49

Borderon JC, Laugier J, Gold F (1978) Essai d'etablissement d'une souche d'Escherichia coli sensible aux antibiotiques dans l'intestin du nouveau-ne. Ann Microbiol 129B:581-596

Calvo FB, Galindo AC, Aspres NB (1992) Iron status in exclusively breast-fed infants. Pediatrics 90:375-9

Cohendy M (1906) Description du ferment lactique puissant capable de s'aclimater dans l'intestine de l'homme. Comptes Rendus de la Soc.de Biol. Paris 60:558

Cook LN, Davis RS, Stover BH (1980) Outbreak of amikacin-resistant enterobacteriaceae in an intensive care nursery. Pediatrics 65:264-268

Duval-Iflah Y, Chappuis JP, Ducluzea R, Raibaud P (1983) Intraspecific interactions between Escherichia coli strains in human newborns and gnotobiotic mice and piglets. Prog.Food Nutr.Sci 7:107-116

Duval-Iflah Y, Ouriet MF, Moreau C, Daniel N, Gabilan JC, Raibaud P (1982) Implantation precoce d'une souche d'Escherichia coli dans l'intestin de nouveau-nes humains: effets de barriere vis-à-vis de souches de *E. coli* antibioresistantes. Ann Microbiol133:393- 408

Elias MF, Nicolson NA, Bora C, Johnston J (1986) Sleep/wake patterns of breast-fed infants in the first 2 years of life. Pediatrics 77:322-329

Fuller R (1989) Probiotics in man and animals: A review J Appl Bacteriol 66:365-78

Gillespie WA, Simpson K, Tozer RC (1958) Staphylococcal infection in a maternity hospital Lancet ii 1075-1080

Goldin BR, Gorbach SL, Saxelin M, Barakat S, Gualtieri L, Salminen S (1992) Survival of Lactobacillus species (strain GG) in human gastrointestinal tract. Digest Dis and Scie 37(1):121-128

Goldmann DA (1988) The bacterial flora of neonates in intensive care-monitoring and manipulation. J Hosp Infect 11:(Supplement A)340-351

Hall MA (1990) Antibiotic policies. Infection in the Newborn John Wiley and Son 11:127-143

Hall MA, Cole CB, Smith SL, Fuller R, Rolles CJ (1990) Factors influencing the presence of faecal lactobacilli in early infancy. Arch Dis Child 65(2):185-8

Herter CA (1897) Certain relations between bacterial activity in the intestine and the indican in the urine. BMJ Dec 25th:1847

Hotto M et al. (1987) Clinical effects of Bifidobacterium preparations on Pediatric Intractable diarrhea. Keio J Med 36(3):298-314

Houck PW, Nelson JD, Kay JL (1972) Fatal septicaemia due to *staphylococcus aureus* 502A. Amer J Dis Child 123:45-8

Isolauri E, Juntunen M, Rautanen T, Sillanaukee P, Kiovula T (1991) A human lactobacillus strain (*lactobacillus GG*) promotes recovery from acute diarrhea in children. Pediatrics 88:90-97

Kershaw GF, Luscombe JR, Cole DJA (1966) Lactic acid and sodium acrylate: effect on growth rate and bacterial flora in the intestines of weaned pigs. Veterinary Rec 79:296

Light IJ, Sutherland JM, Schott JE (1965) Control of a staphylococcal outbreak in a nursery: Use of bacterial interference. JAMA 193:699-704

Light IJ, Walton RL, Sutherland JM, et al. (1967) Use of bacterial interference to control a staphylococcal nursery outbreak. Amer J Dis Child 113:291-300

McNinch AW, Tripp JH (1991) Haemorrhagic disease of the newborn in the British Isles: two year prospective study. BMJ 303:1106-9

Metchnikoff E (1894) Recherches sur le cholera et les vibrions.IV. Sur l'immunité et la receptivité vis-à-vis du cholera intestinal. Ann.Inst.Pasteur (Paris) 8:529-589

Metchnikoff E (1907) The Prolongation of Life. G New York G.P. Putnams Sons

Mollgaard H (1946) On phytic acid, its importance in metabolism and its enzymic cleavage in bread supplemented with calcium. V. Experimental evidence of the beneficial effect of organic oxyacid on the absorption of $Ca^{++}$ and phosphorus from rations containing phytic acid. Biochem J 40:589

Pearce JL, Hamilton JR (1974) Controlled trial of orally administered lactobacilli in acute infantile diarrhea. J Pediatr 84:261-2

Peerson JM, Lonnerdal B, Dewey KG, Heining MJ, Nonurnsen LA (1992) Growth of breast-fed and formula-fed infants from 0 to18 months: the DARLING study. Pediatrics 89:1035-41

Poisson DM, Borderon JC, Amorim-Sena JC, Laugier J (1986) Evolution of the barrier effects against an exogenous drug-sensitive Escherichia coli strain after single or repeated oral administration to newborns and infants aged up to three months admitted to an intensive-care unit. Biol Neonate 49:1-7

Rastegar Lari A, Gold LF, Borderon JC, Laugier J, Lafont J-P (1990) Implantation and *in vivo* antagonistic effects of antibiotic-susceptible Escherichia Coli strains administered to premature newborns. Biol Neonate 56:73-78

Reuman PD, Duckworth DH, Smith KL, Kagan R, Bucciareli RL, Ayoub EM (1986) Lack of effect of lactobacillus on gastrointestinal bacterial colonization in premature infants. Pediatr Infect Dis 5:663-8

Robinson EL, Thompson WL (1952) Effect on weight gain of the addition of *Lactobacillus acidophilus* to the formula of newborn infants. J Pediatr 41:395-8

Sakata H, Yoshioka H, Fujita K (1985) Development of the intestinal flora in very low birth weight infants compared to normal full-term newborns. Eur J Pediatr 144:186-90

Shinefield HR, Ribble JC, Boris M (1971) Bacterial interference between strains of staphylococci aureus, 1960 to 1970. Amer J Dis Child 121:148-152

Shinefield HR, Ribble JC, Bors M et al. (1963) Bacterial interference. Amer J Dis Child 105:646-691

Silva M, Jacobus NV, Deneke C, Gorbach SL (1987) Antimicrobial substance from a human lactobacillus strain. Antimicrob Agents Chemother 31:1231-3

Sprunt K, Leidy G, Redman W (1978) Abnormal colonisation of neonates in an intensive care unit: means of identifying neonates at risk of infection. Pediatr Res 12:998

Sprunt K, Leidy G, Redman W (1980) Abnormal colonisation of neonates in an ICU: conversion to normal colonisation by pharyngeal implantation of alpha haemolytic streptococcus strain 215. Pediatr Res 14:308-13

Tissier H (1900) Rescherches sur la flore intestinale des nourrissons (etat normal et pathologique), These de Paris, Libraire Maloine

Tomic-Karovic K, Fanjek JJ (1962) Acidophilus milk in therapy of infantile diarrhea caused by pathogenic Escherichia coli. Ann Pediatr 199:625-634

Vicek A, Kneifl J (1964) On experiments on normalization of disordered intestinal flora in infants.II. Clinical experiments.I. Omniflora in premature infants. Z. Kinderheilk 89:155-159

Chapter 10

# Commercial Aspects of Formulation, Production and Marketing of Probiotic Products

S. Laulund

The most common use of probiotics is as food in the form of fermented milk products. This is illustrated in the number of review articles reporting scientific and clinical research with fermented milk products compared with the very limited number of publications reporting similar research results with probiotics marketed as pharmaceuticals (Gilliland 1989, Laroia and Martin 1990, Robinson 1991).

Fermented milk products have been defined by the International Dairy Federation (IDF) as a "milk product prepared from milk, skimmed or not, with specific cultures; the microflora is kept alive until sale to the consumer" (IDF 1988). As a consequence all transportation and storage has to be made at low temperatures and the shelf life is limited to a few weeks. This is not unusual for food products as a whole, and compared with other dairy products like milk the shelf life is increased. The logistics of handling this problem is overcome by efficient transportation, daily supply to supermarkets and storage in refrigerators both in the supermarkets and by the consumer. As a consequence it might seem unnecessary to produce freeze-dried probiotics in a similar way as pharmaceuticals, but these formulations offer a number of advantages, for example preparations make it possible for the manufacturers to market their products world–wide as the probiotics can have a shelf life of several years. The sale of probiotics is based on expectations of a curative effect and in many cases as an alternative to more conventional pharmaceutical preparations. The consumers are more inclined to accept the fact that probiotics have an effect when they look like medicine. By means of a special formulation it is also possible to ensure good bacterial survival in the gastric environment. Pharmaceutical preparations make it possible for the user to consume large quantities of viable lactic acid bacteria without necessarily being hungry.

**Fig. 10.1.** A situation to be avoided.

A clinical trial performed during a travel in Egypt (Black *et al.* 1989) proved that a pharmaceutical preparation of a probiotic product (Trevis®) containing four specific bacteria strains had a statistically significant prophylactic effect on traveller's diarrhoea ($p = 0.019$). Fermented milk products containing the same strains may have the same effect. However, it is somewhat difficult to consume 10–20 liter of, for example yoghurt, during a two weeks' period to achieve the same effect. Finally, a probiotic effect from a certain strain does not necessarily mean that dairy products fermented with this strain results in an acceptable foodstuff.

## Formulation of Probiotics

Shortly after Metschnikoff published his advocacy of fermented milk products as beneficial to health (Metschnikoff 1908), the formulation of probiotics as pharmaceuticals started. As early as 1913 a successful treatment of 116 out of 117 cases of infantile diarrhoea with a *Lactobacillus bulgaricus* preparation was reported (Clock 1913).

In Denmark the tablet product, Paraghurt®, (Leo Laboratories) has been on the market since 1922. Several reformulations of the product have changed it from a tablet made of dried yoghurt - in 1929 improved by adding chocolate and coated with paraffin wax (Schroeder 1945) - into the product of today containing freeze–dried *Enterococcus faecium* only, and now without any coating.

It is estimated there must be several thousands of probiotic preparations (tablets, soft and hard gelatine capsules, powders, liquids, pastes, etc.) on the world market.

A microbiological examination of 27 commercial *Lactobacillus acidophilus* preparations performed at The University of Wyoming approximately ten years ago revealed that many of the preparations did not fulfil the declarations on the products.

The findings were: too low a number of viable cells, no viable cells at all, other *Lactobacilli* than *L. acidophilus* and other organisms than lactobacilli, including coliforms and gram–negative rods (Brennan *et al.* 1983).

Within animal probiotic products a similar low quality was found by Gilliland (Gilliland 1981). Only a few correct results were obtained by National Bacteriological Laboratory in Stockholm when they analysed seven probiotic products in a very recent study (CE Nord 1990, personal communication). It seems as the majority of manufacturers have not improved their product quality in the last ten years.

To name all the above mentioned preparations as pharmaceuticals is very unfair to the pharmaceutical industry. In most countries these preparations are marketed without submitting documentation to drug regulating authorities for approval - unlike ethical drugs - with no rules for standards and no official control of quality.

I believe that the reason why the average standard of the so–called pharmaceutical probiotic products is lower than could be expected is that it is very difficult to produce high–quality products and not because the manufacturers just do not care. Poor–quality products do not contribute to a respectable image of probiotics. I am pleased to see that scientists working seriously with probiotics have recently also encouraged the manufacturers to improve the quality of the probiotic preparations. As a beginning, it has been suggested to apply standards for quality control (Fuller 1990), and I am sure that in the near future we shall see an improvement. The health authorities are tightening up the legislation in this field, and in the European Community a common approval system will soon be introduced. This will most likely result in an improved average standard since the number of products will decrease when the low–quality products have been banned.

## Pharmaceutical Formulation of Probiotics

Before developing a pharmaceutical formulation of a probiotic strain many factors have to be considered in order to obtain a high–quality product. Later in the development process, i.e. when the product is to be scaled-up to production scale, new problems very often arise; problems caused by the fact that what seemed to be the optimum formulation appears to be either impossible or uneconomic in this phase. I have identified several important areas.

## Strain Selection

First it is essential to have the appropriate strain or strains. The selection and choice depend on the therapeutical effect required from the product.

When considering the lactic acid bacteria as a whole the range of therapeutical effects possible is extensive.

Review articles mention properties such as nutritional benefits and vitamin synthesis. Treatment of a variety of disorders including colitis, constipation, diarrhoea, flatulence, gastric acidity, gastroenteritis, gingivitis, hypercholesterolemia, hepetic encephalopathy, tumourigenesis, vaginitis and lactose intolerance is also mentioned. Prophylactic use against traveller's diarrhoea and recolonisation of the intestine after antibiotic treatment are also reported (Gorbach 1990).

Some of these properties are based upon sounder scientific proof (Lidbeck *et al.* 1987) than others. Besides, for some of them the *in vitro* results are more convincing than the *in vivo* results.

With such a wide range of possibilities it is clear that strain selection involves a great deal of scientific study. A more detailed description of strain selection is given in chapter 6.

## Number of Viable Cells to Obtain the Required Effect

The procedure for finding the correct therapeutic dose are called dose response studies. At suitable intervals the chosen strain(s) is(are) administered in an increasing concentration to a number of volunteers. This procedure is continued till the amount of viable cells needed to achieve and maintain the required effect has been obtained.

There are few references to *per os* dose response studies with lactic acid bacteria (Friis–Møller and Hey 1983; Saxelin *et al.* 1991).

Only the latter study is a correct study in the pharmacologic sense. They started with a dose of $1.5 \times 10^6$ and stated that freeze–dried powder of *Lactobacillus GG* administered in a dose of $10^{10}$ CFU and $10^{11}$ CFU colonised all volunteers. The difference in the degree of colonisation at the two doses was in proportion to the increase in dose.

An opinion to the therapeutical level and the number of viable cells needed to obtain the required effect is not included in this study. About one year earlier than the above trial the same group of researchers published partly positive results with prophylactic treatment of traveller's diarrhoea by administration of a dose of $2 \times 10^9$ CFU of *Lactobacillus GG* only (Oksanen *et al.* 1990). This may be the explanation of the dilemma they must have been in when they were to conclude what they considered a therapeutic dose. They had earlier observed an effect with a dose which later on proved to be unable to colonise the intestine.

It is important to consider how great indicative importance can be attached to qualitative and quantitative findings made in the faecal flora as regards the possible effect at a higher level of the intestinal system.

Traveller's diarrhoea, which is recognised as a major cause of acute diarrhoeal disease throughout the major part of the world is caused by enterotoxin–producing *Escherichia coli* (ETEC). ETEC have the ability both of adhering to small–bowel epithelial cells and of producing one or more diarrhoeagenic toxins (Carpenter 1983). Therefore, it is most likely that the intake of a specific dose of a probiotic preparation may reach the small intestine and have a beneficial effect (for instance against ETEC) without being traceable in the subsequent faecal flora. Both with regard to quality and quantity there is a significant difference in the composition of the flora from the small intestine to the large intestine. It is in fact very logical that $10^9$ viable cells are capable of exercising an effect on the environment of the small intestine, whereas a concentration one or two logarithms higher is needed before it can be detected in the faeces.

I consider that a dose response study with a probiotic preparation, where the registration of the probiotic strain in faeces is used as an indicator of the obtained therapeutical level, is not necessarily a correct indicator of the obtained therapeutical effect against a large number of the diseases mentioned earlier. I think that the most adequate description of the correct parameter in each individual case at the moment is: "We simply do not know".

The most frequently used method for the determination of dose level today, namely trial and error, will without doubt be the prevailing method for some years ahead. Since adverse reactions with high doses of lactic acid bacteria are very rare and since they are completely harmless the trial and error method is not irresponsible.

## Production of Freeze–Dried Lactic Acid Bacteria

For the production of a high–dose preparation (> $10^9$ per unit), freeze–dried lactic acid bacteria of minimum $10^{10}$ – $10^{11}$ CFU/g are necessary. Such a production requires great skill from the manufacturer in the fermentation and freeze–drying of the bacteria. Only a few manufacturers of probiotic preparations produce bacteria themselves. Most manufacturers buy the bacteria as concentrates.

The therapeutical effect of probiotics is in many cases dependent on a specific strain. In addition trials have been made which describe that the obtained specific properties, e.g. adhesion to epithel cells, are completely dependent on the conditions and fermentation media (Conway *et al.* 1987). The knowledge recently acquired regarding the importance of these special probiotic strains and the importance of the technical fermentation conditions under which they are to be developed presupposes a good cooperation between the manufacturer of the raw materials and the manufacturer of the finished products.

The stability of freeze–dried lactic acid bacteria partly depends on type and strain, but the production procedure also has a crucial influence. The influential parameters can be optimised during the fermentation, but also the composition of added cryoprotectants and the freezing process itself have a crucial influence on the result obtained.

The maintenance of stability of the concentrated bacteria product merely depends on the treatment of it during the further manufacturing and handling. Freeze–dried lactic acid bacteria are very sensitive to humidity. As they are also very hygroscopic, even a short storage under unfavourable conditions may result in a rapid loss in viable cells. If lactic acid bacteria have been exposed to more than 20% – 25% RH the decomposition process cannot be stopped just by removing the bacteria from the humidity or by providing them with moisture–proof packing as an increased water activity remains in the product. A new drying process is the only way to stop the accelerating deterioration, but still the remaining lactic acid bacteria cannot be expected to have the original stability.

Another factor which may influence the number of viable cells is increasing temperature. But if the lactic acid bacteria have for a period of time been kept at temperatures higher than 22–24°C, the increasing deterioration of the viable cells caused thereby can be stopped simply by reducing the temperature.

If the lactic acid bacteria are exposed to increased humidity and increased temperature at the same time, a synergetic effect on the decomposition process will occur.

## Choice of Dosage Form

When the choice of the therapeutic dose has been made, the required quantity of viable cells has to be formulated in a suitable single–dose preparation. During this process the sensitivity of the lactic acid bacteria to e.g. heat, humidity, pressure and low pH must be taken into account. If the application is to be *per os* the administration form must not be unpleasant. Furthermore, the formulation has to protect against the low pH of the gastric juices and at the same time the disintegration time of the formulation has to be short in order to release the bacteria at the very beginning of the intestinal system. After release in the intestine, the lactic acid bacteria undergo rehydration and start to multiply. To help the bacteria in this starting period it is a good idea to add excipients with substrate functions to the formulation.

In connection with formulations for vaginal applications it has to be taken into account that the factors of importance to the disintegration are different from the factors relating to the intestinal system; at the same time formulations causing discharge have to be avoided.

Probiotic preparations are mainly formulated as tablets and hard gelatine capsules packed in bottles or blister sheets or as powder and granulate packed in bottles or as single–dose sachets.

There are advantages as well disadvantages in connection with all formulation types dependent on whether they are considered from a production or an efficiency point of view.

## Tablets

Tablets are a frequently used formulation of a solid dosage form, since they can be produced at a high speed and consequently at low prices. The shape and the surface are suitable for a subsequent coating for the protection of the lactic acid bacteria partly against humidity and partly against the acid environment of the stomach. However, the lactic acid bacteria are not good at tolerating the pressure caused by the compression into tablets.

The quantity of literature describing production and formulation methods which ensure a minimum reduction of the number of viable lactic acid bacteria in tablet preparations is very limited and obsolete (Gumma and Mirimanoff 1971 and Gumma *et al.* 1972). However, literature dealing with the opposite problems that raw materials, utensils and the production process may from time to time cause microbial contamination of the finished product is available. Therefore, trials have been made in order to find the factors of importance to the reduction in number of micro–organisms and to find out how to utilise them during the compression. The results from these trials present a survey of the mechanisms to be avoided during the compression in order to maintain a high cell count when producing a tablet preparation with lactic acid bacteria.

There is no general agreement as to the mechanisms causing harm to the micro–organisms. The suggested mechanisms are as follows:

i The increasing pressure during compression (up to 100–300 $MNm^{-2}$)
ii Shearing forces, i.e. shearing of the particles during the compression
iii General increase in temperature and formation of hot spots.

In research trials Yanagita *et al.* have suggested that shearing forces are the primary mechanism and showed a linear relation between the size of the micro–organism and mortality. Small cells survive better than large cells under the same pressure. At increasing temperatures the number of surviving micro–organisms falls, and Yanagita *et al.* consequently concluded that the heat generated during the compression contributed to cell destruction (Yanagita *et al.* 1978).

Plumpton *et al.* also concluded that shearing forces are the primary cause of cell destruction. They performed trials with different particle sizes of yeast cells and concluded that since the destruction curves vary in relation to the size of the yeast cells, inactivation cannot solely be due to compression (Plumpton *et al.* 1986b). According to the same group of researchers it appears that during compression the temperature of the tablets as a whole is increased by only 5–20°C, dependent on the formulation (Plumpton *et al.* 1986). Consequently, they do not consider it very likely that this factor contributes substantially to the destruction of micro–organisms.

It is generally agreed that the higher the punch pressure, the higher destruction of cells; whereas there are conflicting mathematical/physical explanations of this relation. Fassihi *et al.* have proven that for direct compression mixes which deformate by fracturing there is a linear relationship between the logarithm of percentage survival and applied pressure, which can be designated mathematically by a first order relationship (Fassihi *et al.* 1977).

From a later study it appears that the same does not apply to excipients compressed by means of plastic deformation (Fassihi and Parker 1987). Contrary to this Plumpton *et al.* are of the opinion that there is a linear relation between the destruction of cells and the increased pressure for plastic deformation excipients. However, they were not able to describe this statement by means of an equation (Plumpton *et al.* 1986a).

The differences of opinion as to the importance of the above factors to the survival of micro–organisms in tablets is due to the fact that the factors are affected by a number of other parameters, such as:

i Size of the micro–organisms
ii Particle sizes of the excipients
iii Compression speed
iv Dwell–time (the period of time within which the maximum pressure on the tablets is kept)
v The shape of the punches.

## Size of Micro–Organisms and Excipients

Results of trials with different sized cells show that small organisms have increased chance of survival. However, the particle size of the excipients is also important. The conclusion is that the smaller the micro–organism and the larger the particle size of the excipients, the better the survival of micro–organisms during compression will be. The explanation is that the micro–organisms are able to "hide" in spaces which are formed when using excipients with large particle size. With increasing particle size the number and the size of void spaces will increase, and the micro–organisms will thus escape mechanical damage during the compression and consequently have better survival (Plumpton *et al.* 1986b).

## Compression Speed and Dwell–Time

The dwell–time, which is the period (0.1–0.05 s) within which the maximum pressure on the tablet is kept during the compression, depends on the type of machine used. Generally, eccentric tabletting machines have a longer dwell–time than rotary tabletting machines, but naturally there are variations within these two machine types. The dwell–time also depends on the compression speed - an increased speed will of course result in a shorter dwell–time.

From trials made with different dwell–times it appears that the harmful effect is increased by the length of the dwell–time. This fact stands clearly out when the compression pressure is also high (Plumpton *et al.* 1986).

## Punch Shape

The shape of the punch used for the compression is of importance to the pressure distribution on the compressed tablet (Jacob and Hüttenrauch 1982). At the edge and also in the centre of the tablets compressed by flat punches, there are sections where the pressure, and consequently the degree of compaction and density, is higher than on the remaining parts of the tablets. When compression is made by means of concave punches the pressure distribution on the tablets depends on the concavity of the punches. In tablets compressed by means of slightly concave punches the pressure in the centre and the pressure differences throughout the tablets are larger than in tablets compressed by means of flat punches. On the other hand, when more concave punches are used the pressure homogeneity is increased, i.e. the pressure differences throughout the tablets become smaller. If greatly concaved punches are used, the stress throughout the tablets is increased due to the differences in pressure. The degree of the pressure homogeneity obtained for each of the mentioned types of punches is furthermore dependent on the machine type used for the compression. During compression on an eccentric tabletting machine the upper punch moves downwards and thus presses the powder against the lower punch and the sides, the positions of which are not changed during the compression. On a rotary tabletting machine the upper and the lower punches move against each other during the compression phase. This means that the pressure distribution on the tablet mass will be more homogeneous, irrespective of the type of punch.

Tabletting of lactic acid bacteria is considered the most complicated form of formulation. The required increased costs for the development of an optimum product must be paid by the financial advantage obtained from producing this type of product.

In order to take the many above factors into account in the best and easiest way, tests by means of a computer–operated compression simulator are to be performed. By means of the simulator the compression of a few tablets can be estimated under exactly the same conditions as in a large industrial tabletting machine. At a very early stage of the development of a new tablet it can be tested how the choice of excipients and various processes during the production affects the properties of the tablet. Today these simulators are only available in large pharmaceutical companies and a few universities, and it will be some years before they are in common use.

## Hard Gelatine Capsules

A hard gelatine capsule consists of two separate parts, each a semi–closed cylinder in shape. One of them is called the cap and has a slightly larger diameter than the other, which is called the body and is longer. The cap fits the body tightly, and put together the two cylinders form a sealed unit (Jones 1987). Most probiotic preparations on the market today are available in formulations as hard gelatine capsules. Due to their oblong shape hard gelatine capsules are easily swallowed. However, the large number of probiotic products in capsules is also due to the fact that they can be filled with powder which is easily produced without losing viable cells. In addition, when closed the capsule is not distinguishable from an ethical drug prescribed by a physician. Of course, the fact that the probiotic products in hard gelatine capsules are easily produced does not necessarily mean that they are of low quality. To enhance the resistance of the lactic acid bacteria against humidity and gastric acid, the filled capsules can be enteric film-coated in a fluid-bed system. Another way of improving the capsule formulations is to fill the capsule with lactic acid bacteria in the form of protecting granulate.

The semi-solid matrix formulation technique represents a recent new approach to the filling of hard gelatine capsules. An active ingredient is mixed into a liquefied semi-solid base by the application of heat or shear stress. The liquid is filled into the gelatine body and gets solidified.

This incorporation will result in increased stability of a moisture susceptible material. Controlled release can also be achieved by means of a semi-solid matrix (Bowtle *et al.* 1986). A formulation of lactic acid bacteria products with the use of this technique might be very useful.

## Powder

From a production point of view a probiotic product in the form of a powder is the optimum formulation. Only the choice of excipients and a mixing are needed for a product to be ready for packing.

As regards the effect it is very likely that only a very small quantity of the added lactic acid bacteria will survive the passage through the stomach since the buffer capacity of the powder is very limited. Pettersson *et al.* showed that four strains of lactic acid bacteria were considerably harmed by incubation with gastric juices (Pettersson *et al.* 1983a). In a later study, the strain with the best survival percentage administered as a fermented product passed the stomach with only1.3% viable cells (Pettersson *et al.* 1983b). This means that even a strain selected due to its high acid tolerance will undergo a considerable reduction when passing through the acid environment of the stomach. Financial and practical considerations will be decisive in the choice of a probiotic powder product, of which large quantities need to be consumed, and an alternative formulation.

## Granulate

A granulate is a product consisting of particles from approx. 0.5 mm up to 1 mm – 2 mm. These granulate grains consist of active ingredients (lactic acid bacteria) and excipients held together by a binder. Binders used in connection with granulation are usually dissolved in water or an organic solvent and thereafter added to or sprayed on the powder which is thus made into granulate and then dried. As the use of water as well as solvents is harmful to freeze-dried lactic acid bacteria other and less traditional methods must be used for granulation.

By adding excipients and water-soluble binders to the lactic acid bacteria before freeze-drying the harmful consequences can be avoided, and with the right treatment a granulated product is obtained after freeze-drying. If the granulate has to be made of

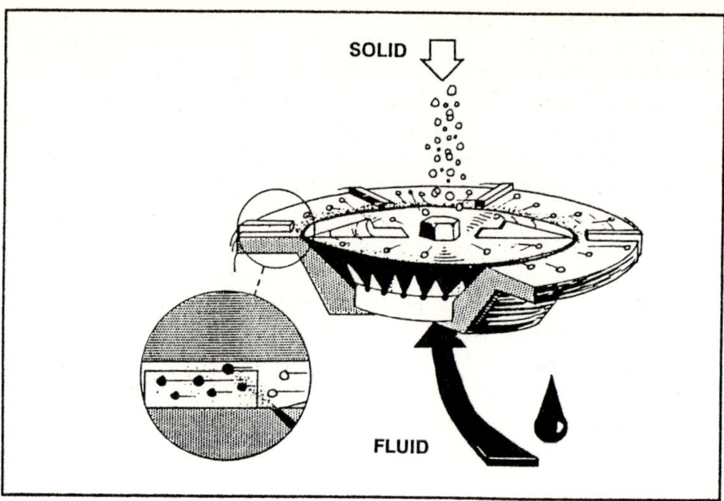


Fig. 10.2. The turbine in a melt-granulation machine.

freeze–dried lactic acid bacteria, melt granulate can be made. The process is as follows: Lipid, wax or a carbohydrate is melted and the now liquid binder is sprayed on or mixed with a powder consisting of lactic acid bacteria and a possible excipient. The powder particles are bound together into a granulate when the binder solidifies.

A modern apparatus capable of performing a melt granulation has been developed and patented by a Swedish company (USA Pat. No 4690834, EPO Applic. No 86850268.3). The main element of the equipment is a turbine. (See Fig. 10.2)

A solid or highly viscous material is liquefied by heating in a melt vessel. The melted material is then pumped into the turbine and instantly atomised. The ensuing mist of fluid material envelopes the solid material that is introduced into the turbine via a separate tube. The material is kept in the turbine chamber for only a fraction of a second, and the coated or granulated material is ejected into a cooling zone.

There are further aspects in the manufacturing of granulate. A granulate is easier to handle during the dosing of the packages and more pleasant to consume than a powder. Binders and excipients may have a protecting effect on lactic acid bacteria against humidity during handling and storage, and they may protect against gastric juices when consumed and thus result in a controlled–release or modified–release product. These properties can with advantage be utilised by filling granulate into hard gelatine capsules or by using the granulate for tablet production.

## Manufacturing Conditions

Various formulation methods have been described. Whatever formulation is chosen, moisture must be prevented from reaching the dry hygroscopic freeze–dried lactic acid bacteria during the process.

Similar problems occur in the production of effervescent tablets and Sendall *et al.* have suggested a method to obtain the best possible conditions in that case. With minor adjustments they can be applied to the production of probiotics.

Moisture ingress is best prevented by controlling the temperature and humidity of the air in all areas where the processing takes place. A relative humidity of 10% – 15% and a temperature of 18°C – 21°C is preferred. A range of equipment is available to provide such conditions but the operation costs are usually high and, therefore, the controlled environment must be kept as small as possible. Air at the required humidity is easily obtained by evaporation of moisture. However, for operator comfort it is usually preferable to provide high humidity in working areas. To balance between both conditions, isolation of controlled humidity areas can be obtained by careful balancing of air supplies and suitable airlocks.

This provision of controlled humidity environments may require special construction materials and surface finishes since conventional materials and adhesives may warp or crack when continuously exposed to low humidity conditions (Sendall *et al.* 1983).

## Packaging

After manufacturing it becomes necessary to prevent deterioration of the formulated probiotic products caused by atmospheric moisture. Sendall *et al.* have also made a number of good suggestions (Sendall *et al.* 1983) on this subject.

One solution is to feed the formulated probiotic product directly into the packing line. However, expensive wastage of product and packing materials may be the consequence if the product does not comply with quality control specifications.

It is more usual to bulk store in conventional containers lined with aluminium foil or a double layer of polythene and containing silica gel sachets. All packing operations should be carried out under controlled humidity conditions. Various packs have been designed to combat the progressive ingress of moist air as the contents are consumed.

For products, which will be consumed shortly after first opening of the pack, a standard glass bottle fitted with a cap of low permeability is satisfactory. A foil cap liner or membrane seal may assist in prolonging the shelf life of unopened packs, while a desiccant canister or sachet should be included to prevent moisture uptake, once the pack is open. This type of packaging relies on original pack dispensing, secure replacement of the closure by the user, and retention of the desiccant in the pack until empty. Plastic containers of low moisture permeability may also be used.

Other widely used methods for preventing moisture ingress are columnar tubes of aluminium, tinplate, glass or plastic equipped with screw, pilfer–proof or plug caps containing silica gel, thus ensuring that the desiccant is always replaced and is close to the weakest point in the package.

A preferred packaging method for tablets and capsules is foil strip or blister packing since only one dose is exposed to the air at a time. A double layer of aluminium foil or a single layer together with a clear film of low moisture vapour permeability is necessary to provide a suitable moisture barrier. However, a problem with this packaging may arise: The very low moisture of the probiotic product tends to make the capsules brittle so that they can be difficult to remove without damage.

# Quality Control

Prior to the production of a probiotic product considerable efforts are made by the manufacturer in order to obtain a high–quality product. However, it can be difficult to estimate whether the product complies with his wishes.

At present the requirements for a probiotic product are very few (however, in most cases the regulations govering claims and advertising are very strict). As previously mentioned Fuller has suggested the establishment of standards for quality control (Fuller 1990).

Based upon my own experience, I suggest the following requirements:

i Uniformation of contents/dosage accuracy
ii Microbial contaminants
iii Long–term stability studies
iv Dissolution/Bioavailability
v Therapeutical effect.

Standard requirements for pharmaceutical solid preparations, on four out of the five mentioned subjects (therapeutical effect is excepted), are described in various pharmacopoeia, i.e. European Pharmacopoeia (Ph. Eur. 1991). It would in practise be very easy to apply the requirements for these standards to probiotics.

The limits for dosage accuracy are as follows: 90% to be within ±10% and 100% to be within ±20% of the intended dosage. Taking into consideration variations in microbiological analyses and productions it could be very expensive to fulfil these requirements. A more appropriate demand would be to require a minimum cell count, taking into consideration that we are dealing with substances of low potency as regards adverse reactions. As another exception, the narrow limits for uniformity of contents do not apply to multivitamin and trace-element preparations.

It is of vital importance to a production that both the type and the number of foreign germs in the final probiotic product are minimised. The limits in the European Pharmacopoeia for "microbial contamination of products not required to comply with the test for sterility", i.e. solid dosage as tablets, capsules, etc., can very easily be applied to probiotics. The Pharmacopoeia sets limits for e.g. *E. coli*, *Salmonella*, *Pseudomonas* and *S. aureus* and sets rules for interpretation of the maximum acceptable limits.

There are no common guidelines on long-term stability studies for pharmaceuticals. The conclusion from a recent conference on the subject was: "The discussion... indicated a substantial lack of agreement on this topic". (PMA 1991). Examples: In the United States the room temperature is equal to 15°C - 30°C, whereas it is equal to 1°C - 30°C in Japan. 30°C is a rather high temperature in respect of probiotic products. This means that defined conditions of storage, under which the shelf life is valid, must appear on the product label.

However, whatever condition is chosen an interpretation is needed of what stability is in connection with a probiotic preparation. According to the European Pharmacopoeia a maximum of 10% reduction is allowed during the shelf life period of a pharmaceutical. The application of these limits raises the same problem as described under dosage accuracy, namely accuracy of microbiological analyses; also in this case a minimum limit of cell count could be a solution. Whether such a minimum limit ought to be

10%, 25% or 50% of the declared starting dose is open to discussion. However, considering the existing uncertainty as to the definition of the therapeutical level and dose response results, arguments in favour of an acceptance of 10%, 25% and 50% can be made.

## Dissolution/Bioavailability

The probiotic preparations, which have obtained an increased acid tolerance and a controlled release via the formulation technique, can be classified as modified–release products.

The bioavailability of these products should be tested in clinical trials during the development phase, e.g. by means of samples collected from duodenum via a nasogastric tube in healthy volunteers who had consumed the product.

To control whether the required bioavailability is obtained throughout the production an *in vitro* measure should be included in the quality control. The European Pharmacopoeia dissolution test (with basket stirring element) is a usable standard method for this test. A thorough review of the experience made and the choice of methods for obtaining an optimum physical tool for the control and estimation of biological availability of a drug has been published by Hanson (Hanson 1982).

The last and very important point, which is usually not considered part of the quality control, deals with clinical trials for control of the efficacy. In my opinion the correct demonstration of the efficacy intended is an important indication of quality. Most probiotic products marketed have not been subject to clinical trials at all, and only a minority of the studies lives up to the rules of Good Clinical Practice for trials on medical products (GCP 1990).

Today it is possible to market a probiotic product without any clinical documentation. On the other hand, in most countries it is prohibited to use results of clinical trials in connection with the marketing of a probiotic product. If such information is to be given, the product has to be registered on terms equal to the terms of medical products. This legislation causes problems.

The result of the cost–benefit analysis, which is made by the company marketing the probiotic product, too often favours the model consisting of a quick marketing of a product not subject to an official control. Instead of spending money on thorough development and testing it is spent on effective advertising campaigns. The companies which are spending money on thorough development and clinical trials cannot utilise their knowledge for purposes of information, and if they want to do so, they are forced to apply for a medical registration, which involves considerable costs and several years' delay of marketing.

A change of legislation is needed in many countries to require quality control of the product and the manufacturer and also documentation of therapeutical efficacy. The fulfilment of these requirements should be made on a level that requires a limited quantity of documentation and a short time for obtaining the approval only. Compared with the registration requirements for medical products, an introduction of such easy terms can be justified for product types like probiotics, where the knowledge of non–toxicity is thorough.

## Afterword

"Although used in humans and animals for generations, probiotics have only recently been subject to scientific and clinical research". This statement probably applies to an even higher degree to the so–called pharmaceutical preparations. People working with dairy products have also understood that probiotics in the form of fermented milk products have the advantage over pharmaceuticals that as dairy products they contain nutrients. So, if the expected and claimed effects fail to appear, at least the consumer is satisfied and has had a tasteful meal.

### *Acknowledgements*

I am indebted to Annegrete Tønners, Adrian T. Gillespie and my wife, Lone Brink Laulund, for their thoughtful help and comments – and to Maj Brit Hansen for the drawing.

## References

Black FT, Andersen PL, Ørskov F, Gaarslev K, Laulund S (1989) Prophylactic Efficacy of *Lactobacilli* on Traveler's Diarrhea. Travel Medicine: 333–335

Bowtle WJ, Lucas RA, Barker NJ (1986) Formulation and Process Studies in Semi–Solid Matrix Capsule Technology. 4th International Conference on Pharmaceutical Technology, Paris 5: 80–89

Brennan M, Wanismail B, Ray B (1983) Prevalence of Viable *Lactobacillus acidophilus* in Dried Commercial Products. J Food Prot 46: 887–892

Carpenter CCJ (1983) Acute Infectious Diarrhoeal Disease and Bacterial Food Poisoning. In: Harrison's Principles of Internal Medicine. Tenth Edition. McGraw Hill Book Company, New York pp 885–889

Clock OR (1913) One Hundred Seventeen Cases of Infant Diarrhea. JAMA 61: 164–168

Conway PL, Gorbach SL, Goldin BR (1987) Survival of Lactic Acid Bacteria in the Human Stomach and Adhesion to Intestinal Cells. Journal of Dairy Science 70: 1–12

European Pharmacopoeia Second Edition (1991). Maisonneuve SA, Sainte–Raffine, France

Fassihi AR, Davies PJ, Parker MS (1977) Effect of Punch Pressure on the Survival of Fungal Spores during the Preparation of Tablets from Contaminated Raw Materials. Zbl Pharm 116: 1267–1271

Fassihi AR, Parker MS (1987) Inimical Effects of Compaction Speed on Microorganisms in Powder Systems with Dissimilar Compaction Mechanisms. Journal of Pharmaceutical Sciences 76: 466–470

Friis–Møller A, Hey A (1983) Colonisation of the Intestinal Canal with A *Streptococcus faecium* Preparation (Paraghurt®). Current Therapeutic Research 33: 807–815

Fuller R (1990) Probiotics in Agriculture. AgBiotech News and Information 2: 217–220

Gilliland SE (1981) Enumeration and Identification of *Lactobacillus acidophilus*. Oklahoma Agricultural Experimental Station Miscellaneous Publication 108: 61– 63

Gilliland SE (1989) *Acidophilus* Milk Products: A Review of Potential Benefits to Consumers. J Dairy Sci 72: 2483–2494

Good Clinical Practice for Trials on Medicinal Product in European Community Approved by CPMP July 1990 (1990). Pharmacology and Toxicology 67: 361– 372

Gorbach SL (1990) Lactic Acid Bacteria and Human Health. Annals of Medicine 22: 37–41

Gumma A, Filthuth I, Mirimanoff A (1972) Etude de quelques procédés galeniques appliqués à la thérapeutique de substitution par *Lactobacillus acidophilus*. 2e communication. Pharmaceitica Acta Helvetica 47: 433–437

Gumma A, Mirimanoff A (1971) Etude de quelques procédés galeniques appliqués à la thérapeutique de substitution par *Lactobacillus acidophilus*. 1re communication. Pharmaceutica Acta Helvetica 46: 278–289

Hanson WA (1982) Handbook of Dissolution Testing. Pharmaceutical Technology Publications, Springfield, Oregon pp 1–163

IDF (1988) Fermented Milks – Science and Technology Bulletin No 227. International Dairy Federation, Brussels pp 164

Jacob J, Hüttenrauch R (1982) Abhängigkeit der Preßdruckverteilung von der Tablettengeometrie. Acta Pharmaceutica Technologica 28: 44–52

Jones BE (1987) The History of the Gelatine Capsule. In: Hard Capsules. Development and Technology (Ridgway K Ed) The Pharmaceutical Press, London pp 1–12

Laroia S, Martin JH (1990) Bifidobacteria as Possible Dietary Adjuncts in Cultured Dairy Products – A Review. Cultured Dairy Prod J November 1990: 18–22

Lidbeck A, Gustafsson JÅ, Nord CE (1987) Impact of *Lactobacillus acidophilus* Supplement on the Human Oropharyngeal and Intestinal Flora. Scandinavian Journal of Infectious Diseases 19: 531–537

Metschnikoff E (1908) The Prolongation of Life. GP Putnam's Sons, New York: 161–183

Oksanen PJ, Salminen S, Saxelin M, Hämäläinen P, Ihantola–Vormisto A, Muurasniemi–Isoviita L, Nikkari S, Oksanen T, Pörsti I, Salminen E, Siitonen S, Stuckey H, Toppila A, Vapaatalo H (1990) Prevention of Travellers' Diarrhoea by *Lactobacillus GG*. Annals of Medicine 22: 53–56

Pettersson L, Graf W, Alm L, Lindwall S, Strömberg A (1983a) Survival of *Lactobacillus acidophilus* NCDO 1748 in the Human Gastrointestinal Tract. 1. Incubation with Gastric Juice *in Vitro*. In: Hallgren B Ed. "Nutrition and the Intestinal Flora". XV Symp. Swed. Nutr. Found'n. Almquist and Wicksell, Uppsala pp 123–125

Pettersson L, Graf W, Sewelin U (1983b) Survival of *Lactobacillus acidophilus* NCDO 1748 in the Human Gastrointestinal Tract. 2. Ability to Pass the Stomach and Intestine *in Vivo*. In: Hallgren B Ed. "Nutrition and the Intestinal Flora". XV Symp.Swed. Nutr. Found'n. Almquist and Wicksell, Uppsala pp 127– 130

Pharmaceutical Manufacturers Association's Joint QC–PDS Stability Committee (1991) Room–Temperature Stability Studies: Storage Conditions. Pharmaceutical Technology International 3: 48–51

Plumpton EJ, Fell JT, Gilbert P (1986) The Survival of Microorganisms during Tabletting, the Influence of Compaction Speed and Dwell time. 4th International Conference of Pharmaceutical Technology, Paris 1: 325–332

Plumpton EJ, Gilbert P, Fell JT (1986a) Effect of Spatial Distribution of Contaminant Microorganisms within Tablet Formulations on Subsequent Inactivation through Compaction. International Journal of Pharmaceutics 30: 237–240

Plumpton EJ, Gilbert P, Fell JT (1986b) The Survival of Microorganisms during Tabletting. International Journal of Pharmaceutics 30: 241–246

Robinson RK (1991) Therapeutic Properties of Fermented Milks. Elsevier Applied Food Science Series, London and New York pp 1–185

Saxelin M, Elo S, Salaminen S, Vapaatalo H (1991) Dose Response Colonisation of Faeces after Oral Administration of *Lactobacillus casei* Strain GG. Microbial Ecology in Health and Disease 4: 209–214

Schroeder K (1945) About Treatment of Intestional Disturbances with Yoghurt Bacteria in Form of Tablets (Paraghurt Tablets). Nord Med 30: 1–5

Sendall FEJ, Staniforth JN, Rees JE, Leatham MJ (1983). Effervescent Tablets The Pharmaceutical Journal 230: 289–294

Yanagita T, Mihi T, Sakai T, Horikoshi I (1987) Microbiological Studies on Drugs and Their Raw Materials. 1. Experiments on the Reduction of Microbial Contaminants in Tablet during Processing. Chem Pharm Bull 26: 185–190

# Subject Index